SpringerBriefs in Computer Science

Series Editors
Stan Zdonik
Shashi Shekhar
Jonathan Katz
Xindong Wu
Lakhmi C. Jain
David Padua
Xuemin (Sherman) Shen
Borko Furht
V.S. Subrahmanian
Martial Hebert
Katsushi Ikeuchi
Bruno Siciliano
Sushil Jajodia
Newton Lee

More information about this series at http://www.springer.com/series/10028

Feifei Gao • Chengwen Xing • Gongpu Wang

Channel Estimation for Physical Layer Network Coding Systems

Springer

Feifei Gao
School of Information
  and Science Technology
Tsinghua University
Beijing, China

Chengwen Xing
School of Information
  and Electronics
Beijing Institute of Technology
Beijing, China

Gongpu Wang
School of Computer and Information
  Technology
Beijing Jiaotong University
Beijing, China

ISSN 2191-5768   ISSN 2191-5776 (electronic)
ISBN 978-3-319-11667-9   ISBN 978-3-319-11668-6 (eBook)
DOI 10.1007/978-3-319-11668-6
Springer Cham Heidelberg New York Dordrecht London

Library of Congress Control Number: 2014950886

Printed on acid-free paper

Springer is part of Springer Science+Business Media (www.springer.com)

Preface

Physical layer network coding (PLNC) system builds a simultaneous bi-directional transmission between two communicating terminals via the aid of a relay and is, sometimes, called as bi-directional relay network. Simultaneous transmission is allowed since any terminal can subtract the self-information from signals that are mixed at the relay. Hence the spectral efficiency is almost enhanced twice as compared to the unidirectional relaying. As any other system, PLNC requires channel state information (CSI) in order to realize the data detection as well as other optimal strategies, e.g., power allocation, node selection, beamforming, etc. To acquire accurate CSI, channel estimation via training sequences sent from both terminals serves as a nature choice. However, the bi-directional two-hop nature makes PLNC different not only from the traditional point-to-point system but also from unidirectional relaying system (URS). Hence, the existing channel estimation strategies designed for point-to-point system or URS, if applied to PLNC, would suffer from spectral inefficiency. It is then necessary to re-look into the channel estimation methodology and design the corresponding training sequences that are suitable for PLNC.

The objective of this Springer brief is to present the architectures of the PLNC system and examine recent advances in channel estimation for such a system. The motivations and concepts of PLNC are first explored. Then the challenges of channel estimation as well as other signal processing issues in PLNC are presented. The readers are exposed to the latest channel estimation and training sequence designs for PLNC system under three typical fading scenarios: frequency flat fading, frequency selective fading, and time selective fading. Via estimation theory and optimization theory, the new channel estimation mechanisms in PLNC system are devised to embrace the bi-directional two-hop nature, and the corresponding optimal

training structures are also derived. Numerical results show the effectiveness of the
new estimation strategies and the optimality of the training designs.

Beijing, China   Feifei Gao
Beijing, China   Chengwen Xing
Beijing, China   Gongpu Wang
July 2014

Contents

Acronyms

PLNC	Physical layer network coding
CSI	Channel state information
URS	Uni-directional relaying system
SISO	Single-input single-output
MISO	Multiple-input single-output
MIMO	Multiple-input multiple-output
OFDM	Orthogonal frequency division multiplexing
STC	Space time coding
ML	Maximum likelihood
LS	Least square
MSE	Mean square error
MMSE	Minimum mean square error
LMMSE	Linear minimum mean square error
SNR	Signal-to-noise raio
LMSNR	Linear maximum signal-to-noise ratio
CRLB	Cramér-Rao lower bound
PDF	Probability density function
AF	Amplify-and-forward
AESNR	Average effective SNR
AMSE	Average MSE
SER	Symbol error rate
BER	Bit error rate
PT	Pilot-tone
TDD	Time-division-duplexing
DFT	Discrete Fourier transformation
IDFT	Inverse discrete Fourier transformation
SSA	Simultaneous sign ambiguity
CE-BEM	Complex exponential basis expansion model

Chapter 1
Fundamentals of Physical Layer Network Coding

Abstract Network coding is a beautiful and powerful technique whose emergence invokes a wide range of applications. Along with the development of the theoretical research and wide-spread applications, just like its original butterfly network topology, network coding metamorphoses from a complex and abstruse theory to a widely accepted and natural choice for wireless designs. The mystery and nature behind network coding become more and more attractive and important for wireless researchers. In this very first chapter, we briefly review network coding theory. Specifically, the physical layer network coding (PLNC) is discussed in more detail and understood from three different viewpoints. This way reveals the relationships among PLNC and several existing transmission schemes, which are helpful for us to exploit the performance advantages promised by PLNC.

1.1 Preliminaries of Physical Layer Network Coding

Network coding is a novel, elegant and powerful technique for network communications [1], and has attracted significant interests from several research communities, such as information theory, communication theory and computer science. Basically, network coding has changed the thinking logic about network communications especially for the mutual interference [2]. When the intermediate nodes combine and process the received information streams, the original network coding theory can clearly specify the necessary and sufficient conditions for unicast rate to each receiver with a given rate that also works for multicast at the same rate. The success of network coding makes the researchers believe it is a critical technique for future networks including both wireless and wired networks. Network coding then became a promising and fancy research direction, which is widely accepted as a breakthrough technique to satisfy ever increasing demand of communication rates [3, 4]. More and more applications have been discovered as shown in Fig. 1.1, in which network coding exhibits its great advantages in terms of benefits, complexity, delay and so on [5–7].

Referring to communication networks, the designs for wireless networks are more challenging because of the openness of wireless channels. In wireless networks, when there is more than one terminal simultaneously transmitting signals, the electromagnetic signals are superimposed with each other. As a result,

© The Author(s) 2014   1
F. Gao et al., *Channel Estimation for Physical Layer Network Coding Systems*,
SpringerBriefs in Computer Science, DOI 10.1007/978-3-319-11668-6_1

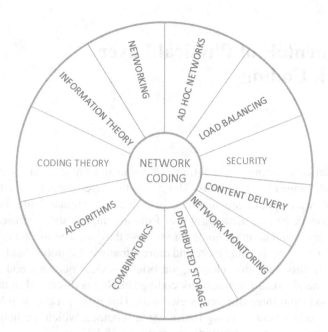

Fig. 1.1 Various application fields of network coding

at any receiver the undesired signals become the interference. More importantly, interference is the most limiting bottleneck for wireless performance. To harness the interference and improve the throughput, the concept of network coding can be extended to physical layer, i.e., the so named physical layer network coding (PLNC) [8]. Comparing to routing strategies, network coding based transmissions, especially physical layer based ones, enjoy much higher throughput.

In the following, a bi-directional relay channel is discussed in detail to illustrate the key idea of PLNC, which is the simplest but most representative example. Specifically, as shown in Fig. 1.2 there are two nodes A and C that wish to exchange messages x_1 and x_2 via a relay node B. The basic routing strategy needs four time slots to exchange information: In the first time slot, node A sends message x_1 to relay B. Then in the second time slot, node B forwards the massage to node C. A similar process is repeated next for node C. On the other side, a network coding based strategy needs three time slots: During the first time slot, node A sends message x_1 to relay C while in the second time slot node C sends message x_2 to relay B. During the third time slot, the relay node B broadcasts the sum of the messages $x_1 \oplus x_2$ to both nodes.

It is well-established that decoding the received signals at the relay is not necessary, and the relay nodes can simply forward the received signals to the desired destinations. Capitalizing this fact, we may ask the relay node B to receive the superposition of the transmitted signals $x_1 + x_2$ from nodes A and B in the first time slot, and then forward it to both nodes A and B in the second time slot. At node A, x_1 is perfectly known and can be removed from the received signal. A similar process

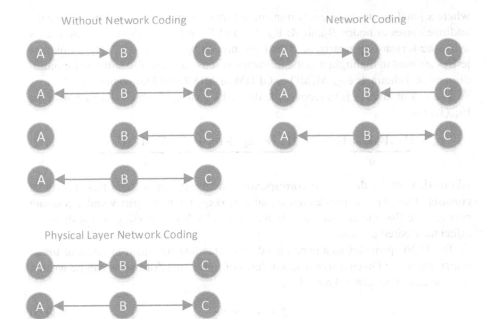

Fig. 1.2 The diagrams of different transmission strategies

also happens at node C. Such the transmission only needs two time slots to complete the information exchange and is known as PLNC.

Besides the previous simplest example, configuration of PLNC can also involve multiple-input multiple-output (MIMO) channels, orthogonal frequency division multiplexing (OFDM) modulations, space time coded (STC) transmission and so on. In most of the existing works, PLNC is understood from information theoretic or at least coding perspective, and there is a rich body of tutorials, surveys or books on these topics. Nevertheless, in this book, we will try to reveal the basic ideas behind PLNC from the signal processing viewpoint. In particular, we will present three aspects of PLNC, i.e., data detection, virtual unidirectional relaying, and soft combination of uplink and downlink.

1.2   Data Detection

Denoting the channel from node X to node Y as $\mathbf{H}_{YX}$ where X and Y can be A, B or C, the received signal at node A is written as[1]

$$\mathbf{r}_A = \mathbf{H}_{AB}\mathbf{F}_B(\mathbf{H}_{BA}\mathbf{F}_A\mathbf{x}_A + \mathbf{H}_{BC}\mathbf{F}_C\mathbf{x}_C) + \mathbf{H}_{BA}\mathbf{F}_B\mathbf{n}_B + \mathbf{n}_A \qquad (1.1)$$

---

[1]Due to symmetry, similar discussions hold for node C and will be omitted here.

where $\mathbf{x}_A$ and $\mathbf{x}_C$ are the signals transmitted from A and C; $\mathbf{n}_B$ and $\mathbf{n}_A$ denote the additive noises at nodes B and A; $\mathbf{F}_A$, $\mathbf{F}_B$ and $\mathbf{F}_C$ represent the precoding matrix at A, the forwarding matrix at B and the precoding matrix at C. Here boldface letters are used to highlight that the following discussions are also suitable for multi-dimensional channels e.g., MIMO or OFDM or MIMO-OFDM channels.

The task at node A is to recover the desired signal $\mathbf{x}_C$ from $\mathbf{r}_A$. Reformulate the Eq. (1.1) as

$$\mathbf{r}_A = \underbrace{\mathbf{H}_{AB}\mathbf{F}_B\mathbf{H}_{BC}\mathbf{F}_C}_{\mathbf{H}}\,\mathbf{x}_C + \underbrace{\mathbf{H}_{AB}\mathbf{F}_B\mathbf{n}_B + \mathbf{n}_A}_{\mathbf{v}} + \underbrace{\mathbf{H}_{AB}\mathbf{F}_B\mathbf{H}_{BA}\mathbf{F}_A\mathbf{x}_A}_{\mathbf{c}}, \qquad (1.2)$$

where $\mathbf{H}$, $\mathbf{v}$, and $\mathbf{c}$ denote the corresponding items. It can be seen that the signal consists of three parts: the desired signal part $\mathbf{H}\mathbf{x}_C$, the noise part $\mathbf{v}$ and a constant part $\mathbf{c}$. Note that the constant part is introduced by $\mathbf{X}_A$, removing which does not affect the desired detection.

The PLNC provides us a generalized understanding for linear systems or linear transformations: Given a signal $\mathbf{x}$, any form of linear transformation can be written into the following general model

$$\mathbf{y} = \mathbf{H}\mathbf{x} + \mathbf{c}, \qquad (1.3)$$

where $\mathbf{H}$ and $\mathbf{c}$ are constant. With some mild conditions satisfied, e.g., dimensionality constraints, the desired signal can still be recovered uniquely. Furthermore, when the unknown noises are taken into account, the signal model becomes

$$\mathbf{y} = \mathbf{H}\mathbf{x} + \mathbf{c} + \mathbf{v}, \qquad (1.4)$$

where $\mathbf{v}$ represents the unknown noise. In practice, $\mathbf{H}$, $\mathbf{c}$ and the covariance of $\mathbf{v}$ are functions of system parameters such as precoding matrices, forwarding matrices and so on. The system parameters should be optimized to achieve better performance. It is interesting that no matter how the system parameters are optimized, there is no performance gain coming from the term c.

For MIMO and OFDM, only $\mathbf{H}$ is a function of system parameters [9], and for uni-directional relaying systems (URS) only $\mathbf{H}$ and the covariance of $\mathbf{v}$ are functions of system parameters, while for PLNC, all terms are functions of system parameters. Hence, PLNC is a specific scheme to build (1.4).

1.3 Virtual Unidirectional Relaying

Before the emergence of bi-directional relaying, the traditional unidirectional strategies have already been well studied. Actually, bi-directional relaying can also be understood from unidirectional relaying. Taking the simplest three-node bi-directional relaying channel as the example, as shown in Fig. 1.3, we see that bi-directional relaying is a special case of the unidirectional relaying with two source-destination pairs. In other words, PLNC can be recognized as virtual URS.

Fig. 1.3 Virtual
unidirectional relaying
system

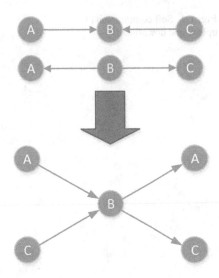

More specifically, in Fig. 1.3 if we use different symbols to denote A and C in the second time slot, the network model becomes a standard unidirectional relaying with two source destination pairs and one relay. The cost of this virtualization is the doubled numbers of source and destination nodes in the equivalent model.

This observation reveals why many algorithms originally designed for URSs can be directly applied to PLNC. For example, the iterative linear minimum mean square error (LMMSE) transceiver designs for unidirectional relaying and bi-directional relaying are exactly the same [10]. The minor difference in the problem formulation may be only the self-interference part, i.e., each receiver needs to remove its own transmitted signals in the first slot.

However, it must be highlighted that compared to URS, bi-directional relaying has some special structures that can be further exploited. It is obvious that if the same frequency band is used, channels in PLNC satisfy $\mathbf{H}_{XY} = \mathbf{H}_{YX}^{\mathrm{T}}$, which allows us to derive much stronger results for PLNC than those for virtualized URS. For example, when all nodes are equipped with multiple antennas, the so obtained channel structures allow us to simultaneously transform the two hop channels into a series of orthogonal eigenmodes. Then the communication designs, such as power allocation, can be significantly simplified.

Moreover, as will be seen in later chapters that the channel estimation of $\mathbf{H}_{XY}$ could be very different in PLNC than in any other linear system.

1.4 Soft Combination of Uplink and Downlink

From another viewpoint, PLNC can be understood as a soft combination of uplink and downlink channels. As shown in Fig. 1.4, two mobile terminals A and C in a cell wish to exchange the information via the base station (BS) B. Without considering

Fig. 1.4 Soft combination of
uplink and downlink

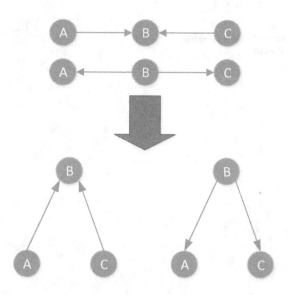

the core networks, the traditional communication will involve two phases i.e., uplink transmission and downlink transmission. In the uplink phase, the two wireless terminals transmit their messages to BS and then BS equalizes the received signals and demodulated them. In the next downlink phase, the messages will be transmitted to the corresponding terminals.

The main disadvantage of this scheme is that there will be some information loss in the message transmission as hard decisions have been made at BS. Especially, in low signal-to-noise ratio (SNR) region the hard decisions will inevitably induce performance loss. With the premise that only linear operations are adopted at BS, the optimal strategy is BS directly forwarding the linearly processed messages to the wireless terminals. Then the communication scheme becomes the same as bi-directional relaying. In other words, the role of BS simply reduces to relaying.

This fact reveals that PLNC system with multiple source-destination pairs will inherit all the technical challenges for the designs in uplink and downlink transmissions. In particular, if the channels are MIMO channels, the bi-directional relaying schemes are soft combinations of uplink and downlink multi-user multi-input multi-output (MU-MIMO) systems. As a result some technical bottlenecks of the designs for bi-directional relaying directly come from the ones for uplink and downlink MU-MIMO systems such as co-channel interference management [11]. On the other hand, another difficulty in the designs for bi-directional relaying comes from relaying nature, e.g., the noise at the relay will be amplified and forwarded to the destination. It can be concluded that in general the problems in bi-directional relaying are definitely more challenging.

1.5 Conclusions

In this chapter, we introduce the background of network coding and discuss the PLNC in depth. We present PLNC from three different perspectives which reveal the relationships between PLNC and several other communications/signal processing problems. These facts can facilitate the readers to understand the nature of PLNC.

References

1. R. Ahlswede, N. Cai, S.-Y. R. Li, and R. W. Yeung. Network information flow. *IEEE Trans. Inf. Theory*, 56(7), pp. 1204–1216, July 2000.
2. S.-Y. R. Li, R. W. Yeung, and N. Cai. Linear network coding. *IEEE Trans. Inf. Theory*, 49(2), pp. 371–381, Feb. 2003.
3. R. W. Yeung, S.-Y. R. Li, N. Cai, and Z. Zhang. Network coding theory. *Foundation and Trends in Communications and Information theory*, 2(4), pp. 241–381, 2005.
4. R. W. Yeung, S.-Y.R. Li, N. Cai, and Z. Zhang. Network coding theory part I: Single source. *Foundation and Trends in Communications and Information Theory*, 2(4), pp 241–329, 2005
5. A. G. Dimakis, V. Prabhakaran, and K. Ramchandran. Network coding for distributed storage systems. in *Proc. of Infocom*, 2007.
6. C. Fragouli and E. Soljanin. Network coding fundamentals. *Foundation and Trends in Networking*, vol. 2, pp. 1–133, 2007.
7. N. Cai and R. W. Yeung. Secure network coding. *IEEE International Symposium on Information Theory*, Lausanne, Switzerland, Jun. 30-Jul. 5, 2002.
8. S. Zhang, S. C. Liew, and P. P. Lam. Hot topic: physical layer network coding. in *Proc. of 12th MobiCom*, pp. 358–365, New York, NY, USA, 2006.
9. D. Tse and P. Viswanath, *Fundamentals of Wireless Communication*. Cambridge University Press. 2005.
10. C. Xing, S. Li, Z. Fei, and J. Kuang. How to understand linear minimum mean square error transceiver design for multiple input multiple output systems from quadratic matrix programming. *IET Communications*, 7(12), pp. 1231–1242, Aug. 2013.
11. E. G. Larsson and P. Stoica, *Space-Time Block Coding for Wireless Communications*. Cambridge University Press, 2003.

Chapter 2
Background on Channel Estimation

Abstract In this chapter, we introduce popular channel estimation approaches in the conventional point-to-point MIMO system, including the maximum likelihood (ML) estimation and minimum mean square error (MMSE) estimation, as well as their counterparts least square (LS) estimation and linear minimum mean square error (LMMSE) estimation. Moreover, the channel estimation for the amplify-and-forward (AF) unidirectional relaying system (URS) is also discussed, which is seen to be quite different from that of the point-to-point system. One should identify the "needed" channel parameters in URS and then carry out the corresponding estimation. Finally, we discuss the differences of the channel estimation in PLNC system and present the new challenges compared to the other systems.

2.1 Channel Estimation in Point-to-Point System

Most communication systems consist of two periods: training period and data transmission period. During the former period, channel is estimated from known symbols, namely the training sequence or pilots; During the latter period, the estimated channel is used to detect the unknown data symbols.

In the point-to-point system, the channel estimation model is usually formulated as

$$\mathbf{y} = \mathbf{Sh} + \mathbf{n}, \tag{2.1}$$

where $\mathbf{y}$ is the received signal during the training period, $\mathbf{S}$ is the matrix that is built from the training sequence, $\mathbf{h}$ is the unknown channel to be estimated, and $\mathbf{n}$ is the unknown noise. Note that the structure of $\mathbf{S}$ could be vary under different configurations.

2.1.1 Estimation of Deterministic Channel

The probability density function (PDF) of $\mathbf{y}$ conditioned on $\mathbf{h}$, i.e., $p(\mathbf{y}|\mathbf{h})$, is also named as the *likelihood function* because it tells us how likely a certain $\mathbf{y}$ is observed with a given $\mathbf{h}$. The ML estimation for $\mathbf{h}$ is then defined as

© The Author(s) 2014
F. Gao et al., *Channel Estimation for Physical Layer Network Coding Systems*,
SpringerBriefs in Computer Science, DOI 10.1007/978-3-319-11668-6_2

$$\hat{\mathbf{h}}_{ML} = \arg\max_{\mathbf{h}} p(\mathbf{y}|\mathbf{h}). \tag{2.2}$$

When noise is Gaussian distributed with covariance matrix $\mathbf{C_n}$, $p(\mathbf{y}|\mathbf{h})$ can be explicitly written as

$$p(\mathbf{y}|\mathbf{h}) = \frac{1}{|\pi \mathbf{C_n}|} \exp\left[-(\mathbf{y} - \mathbf{Sh})^H \mathbf{C_n}^{-1}(\mathbf{y} - \mathbf{Sh})\right]. \tag{2.3}$$

From (2.2), we know the ML estimation of $\mathbf{h}$ can be derived as

$$\hat{\mathbf{h}}_{ML} = \arg\min_{\mathbf{h}} \underbrace{(\mathbf{y} - \mathbf{Sh})^H \mathbf{C_n}^{-1}(\mathbf{y} - \mathbf{Sh})}_{J_{ML}}, \tag{2.4}$$

where J_{ML} is the corresponding cost function. Setting the gradient of J_{ML} with respect to $\mathbf{h}$ as zero, we obtain

$$\hat{\mathbf{h}}_{ML} = (\mathbf{S}^H \mathbf{C_n}^{-1} \mathbf{S})^{-1} \mathbf{S}^H \mathbf{C_n}^{-1} \mathbf{y}. \tag{2.5}$$

The expectation of $\hat{\mathbf{h}}_{ML}$ is

$$E_{\mathbf{y}|\mathbf{h}}\{\hat{\mathbf{h}}_{ML}\} = \mathbf{h}, \tag{2.6}$$

which means that $\hat{\mathbf{h}}_{ML}$ is unbiased, and the covariance matrix of the estimation error vector $\Delta\mathbf{h}_{ML} \triangleq \mathbf{h} - \hat{\mathbf{h}}_{ML}$ can be computed as

$$\mathbf{C}_{\Delta\mathbf{h}_{ML}} = E_{\mathbf{y}|\mathbf{h}}\left\{\Delta\mathbf{h}_{ML}\Delta\mathbf{h}_{ML}^H\right\} = (\mathbf{S}^H \mathbf{C_n}^{-1} \mathbf{S})^{-1}. \tag{2.7}$$

However for a general case, we do not have statistic knowledge of the noise vector $\mathbf{n}$, and hence we could resort to the LS estimation. The principle of LS estimator is to minimize the square norm between the observation $\mathbf{y}$ and the noise-free data, i.e.,

$$\hat{\mathbf{h}}_{LS} = \min_{\mathbf{h}} \underbrace{\|\mathbf{y} - \mathbf{Sh}\|^2}_{J_{LS}}, \tag{2.8}$$

where J_{LS} is the corresponding cost function. Making the derivative of J_{LS} with respect to $\mathbf{h}$ be zero yields LS estimation

$$\hat{\mathbf{h}}_{LS} = (\mathbf{S}^H \mathbf{S})^{-1} \mathbf{S}^H \mathbf{y}. \tag{2.9}$$

Note that when noise $\mathbf{n}$ is Gaussian with $\mathbf{C_n} = \mathbf{I}$, LS estimation coincides with ML estimation.

2.1.2 Estimation of Random Channel

If $\mathbf{h}$ is assumed to be a random vector whose prior knowledge, e.g., PDF or statistics, is known, then one can rely on this prior knowledge to improve the estimation accuracy.

Generally, the optimal estimation is obtained from the MMSE criterion, i.e.,

$$\hat{\mathbf{h}}_{MMSE} = \arg \min_{\mathbf{h}} \underbrace{E_{\mathbf{y},\mathbf{h}}\{\|\hat{\mathbf{h}} - \mathbf{h}\|^2\}}_{J_{MMSE}}, \tag{2.10}$$

where J_{MMSE} is the corresponding cost function. With the equality $p(\mathbf{y}, \mathbf{h}) = p(\mathbf{h}|\mathbf{y}) p(\mathbf{y})$, the cost function J_{MMSE} can be reexpressed as

$$J_{MMSE} = \mathbb{E}_{\mathbf{y}} \left\{ \mathbb{E}_{\mathbf{h}|\mathbf{y}} \left\{ \|\hat{\mathbf{h}} - \mathbf{h}\|^2 \right\} \right\}$$
$$= \int \left[\int \|\hat{\mathbf{h}} - \mathbf{h}\|^2 p(\mathbf{h}|\mathbf{y}) d\mathbf{h} \right] p(\mathbf{y}) d\mathbf{y}. \tag{2.11}$$

Since $p(\mathbf{y}) \geq 0$ for all $\mathbf{y}$, we may simplify the MMSE estimator as

$$\hat{\mathbf{h}}_{MMSE} = E_{\mathbf{h}|\mathbf{y}}\{\mathbf{h}\}. \tag{2.12}$$

If $\mathbf{n}$ and $\mathbf{h}$ are assumed to be joint circularly complex symmetric Gaussian distributed, then the MMSE estimator of $\mathbf{h}$ can be simplified as

$$\hat{\mathbf{h}}_{MMSE} = \mu_{\mathbf{h}} + \mathbf{C}_{\mathbf{h}}\mathbf{S}^H (\mathbf{S}\mathbf{C}_{\mathbf{h}}\mathbf{S}^H + \mathbf{C}_{\mathbf{n}})^{-1}(\mathbf{y} - \mathbf{S}\mu_{\mathbf{h}})$$
$$= \mu_{\mathbf{h}} + (\mathbf{C}_{\mathbf{h}}^{-1} + \mathbf{S}^H \mathbf{C}_{\mathbf{n}}^{-1}\mathbf{S})^{-1}\mathbf{C}_{\mathbf{n}}^{-1}(\mathbf{y} - \mathbf{S}\mu_{\mathbf{h}}), \tag{2.13}$$

where $\mu_{\mathbf{h}}$ and $\mathbf{C}_{\mathbf{h}}$ are the mean and covariance of $\mathbf{h}$. The estimation error $\Delta\mathbf{h}_{MMSE} = \mathbf{h} - \hat{\mathbf{h}}_{MMSE}$ is then also circularly complex symmetric Gaussian distributed with the covariance matrix

$$\mathbf{C}_{\Delta\mathbf{h}_{MMSE}} = \mathbb{E}_{\mathbf{y},\mathbf{h}}\{\Delta\mathbf{h}_{MMSE}(\Delta\mathbf{h}_{MMSE})^H\} = (\mathbf{C}_{\mathbf{h}}^{-1} + \mathbf{S}^H\mathbf{C}_{\mathbf{n}}^{-1}\mathbf{S})^{-1}. \tag{2.14}$$

If only the second order statistics of $\mathbf{n}$ and $\mathbf{h}$ are known, then we may chose to retain the MMSE criterion but constrain the estimator to be linear. The so derived estimator is named as LMMSE estimator, which is formulated as $\hat{\mathbf{h}}_{LMMSE} = \mathbf{b}^H\mathbf{y}$, where the vector $\mathbf{b}$ is achieved through minimizing the following cost function

$$J_{LMMSE} = E_{\mathbf{y},\mathbf{h}}\{\|\mathbf{b}^H\mathbf{y} - \mathbf{h}\|^2\}. \tag{2.15}$$

After some mathematical operations, the LMMSE estimation of $\mathbf{h}$ can be written as

$$\hat{\mathbf{h}}_{LMMSE} = \mu_{\mathbf{h}} + (\mathbf{C}_{\mathbf{h}}^{-1} + \mathbf{S}^H \mathbf{C}_{\mathbf{n}}^{-1} \mathbf{S})^{-1} \mathbf{C}_{\mathbf{n}}^{-1} (\mathbf{y} - \mathbf{S}\mu_{\mathbf{h}}), \tag{2.16}$$

and the covariance matrix for the estimation error vector is

$$\mathbf{C}_{\Delta \mathbf{h}_{LMMSE}} = (\mathbf{C}_{\mathbf{h}}^{-1} + \mathbf{S}^H \mathbf{C}_{\mathbf{n}}^{-1} \mathbf{S})^{-1}. \tag{2.17}$$

Hence, the expression of LMMSE estimator coincides with MMSE estimator (2.13) when $\mathbf{n}$ and $\mathbf{h}$ are jointly Gaussian distributed.

2.2 Channel Estimation in AF Based URS

A typical AF based URS with M randomly placed relay nodes $\mathbb{R}_i$, $i = 1, \ldots, M$, one source node $\mathbb{S}$, and one destination node $\mathbb{D}$ is shown in Fig. 2.1. Compared to the conventional point-to-point system, there are three different types of channel parameters to be estimated: the individual channel $\mathbb{S} \rightarrow \mathbb{R}$ denoted as g_i's, the individual channel $\mathbb{R} \rightarrow \mathbb{D}$ denoted as h_i's, as well as the composite channels $\mathbb{S} \rightarrow \mathbb{R} \rightarrow \mathbb{D}$. Though the ideal channel estimation could target at the two individual channels g_i's and h_i's, [1] has mentioned that in order to achieve the maximum likelihood detection only the composite channels are needed. Interestingly, this idea reminds us that the ultimate purpose of the channel estimation is to realize the data detection, while one would only need those necessary channel information for data detection. In the conventional point-to-point system, such "needed" channel is just the channel between the transceivers, while in URS, one has to figure out what kind of channel is the "needed" one. In order to better illustrate this concept, we present a channel estimation example for URS with multiple relay nodes, where the space time coding (STC) is applied for data transmission.

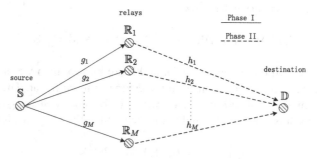

Fig. 2.1 Wireless relay networks with one source, one destination and M relays

2.2.1 Space Time Coding in AF Based URS

Consider the URS shown in Fig. 2.1, where each node has only a single antenna that cannot transmit and receive simultaneously. The channel between each node pair is assumed quasi-stationary with variances σ_{gi}^2 and σ_{hi}^2, respectively. The source node $\mathbb{S}$ sends a signal block $\mathbf{s} = [s_1, \ldots, s_T]^T$ to $\mathbb{D}$ via the aid of relay. The transmission is accomplished by two phases, each containing T consecutive time slots. During Phase I, $\mathbb{S}$ broadcasts the signal $\mathbf{s}$ to all $\mathbb{R}_i$, and the received signal at $\mathbb{R}_i$ is

$$\mathbf{r}_i = g_i \mathbf{s} + \mathbf{n}_{ri}, \tag{2.18}$$

where $\mathbf{n}_{ri}$ is the white complex Gaussian noise at the ith relays. For convenience, all noise variances are assumed as σ_n^2, namely, $\mathbf{n}_{ri} \in \mathcal{CN}(0, \sigma_n^2 \mathbf{I})$. The power constraint of the transmission is $\mathrm{E}\{\mathbf{s}^H \mathbf{s}\} = T P_s$, where P_s is the average transmitting power of the source.

The linear dispersion based STC has been proposed in [2], where $\mathbf{r}_i$ is firstly precoded by a unitary matrix $\mathbf{P}_i$ and is then scaled by a real factor α_i to keep the average power of $\mathbb{R}_i$ as P_{ri}, resulting in

$$\mathbf{t}_i = \alpha_i \mathbf{P}_i \mathbf{r}_i^{(*)}, \tag{2.19}$$

where $(\cdot)^{(*)}$ represents the item itself if the ith relay operates on $\mathbf{r}_i$ whereas represents the conjugate of the item if the ith relay operates on $\mathbf{r}_i^*$. Note that this type of STC, where one relay operates on either $\mathbf{r}_i$ or $\mathbf{r}_i^*$, exclusively, has been adopted in [2, 3]. Moreover, the scaling factor α_i could be chosen as

$$\alpha_i = \sqrt{\frac{P_{ri}}{\sigma_{gi}^2 P_s + \sigma_n^2}} \tag{2.20}$$

to keep the power constraint from the long term point of view. The destination $\mathbb{D}$ in Phase II then receives

$$\mathbf{d}_2 = \sum_{i=1}^{M} h_i \mathbf{t}_i + \mathbf{n}_{d2} = \mathbf{B}\mathbf{\Lambda}\mathbf{w} + \mathbf{n}_d, \tag{2.21}$$

where $\mathbf{n}_{d2} \in \mathcal{CN}(0, \sigma_n^2 \mathbf{I})$ represents the complex white Gaussian noise vector at $\mathbb{D}$ in the second phase, and

$$\mathbf{w} = [w_1, \ldots, w_M]^T, \quad w_i = h_i g_i^{(*)}, \quad \mathbf{\Lambda} = \mathrm{diag}\{\alpha_1, \ldots, \alpha_M\},$$

$$\mathbf{B} = [\mathbf{P}_1 \mathbf{s}_1^{(*)}, \mathbf{P}_2 \mathbf{s}_2^{(*)}, \ldots, \mathbf{P}_M \mathbf{s}_M^{(*)}], \quad \mathbf{n}_d = \sum_{i=1}^{M} h_i \alpha_i \mathbf{P}_i \mathbf{n}_{ri}^{(*)} + \mathbf{n}_{d2}.$$

Note that, by a slight abuse of notation we introduce the notation $\mathbf{s}_i \triangleq \mathbf{s}$ to discriminate the signal forwarded from the ith relay. Furthermore, the covariance of $\mathbf{n}_d$ is computed as

$$\text{Cov}(\mathbf{n}_d | h_i, i = 1, \ldots, M) = \left(\sum_{i=1}^{M} |h_i|^2 \alpha_i^2 + 1 \right) \sigma_n^2 \mathbf{I}, \qquad (2.22)$$

where the property $\mathbf{P}_i \mathbf{P}_i^H = \mathbf{I}$ is utilized.

2.2.2   Channel Estimation

For coherent detection in the AF based URS [2–5], the destination $\mathbb{D}$ performs the maximum likelihood (ML) detection based only on a specific channel realization w_i while treating $\mathbf{n}_d$ as the overall white Gaussian noise. Therefore, the task of the channel estimation focuses only on estimating w_i at $\mathbb{D}$.

Two different channel estimation schemes could be considered. One is to separately estimate g_i, h_i and then construct w_i from $g_i^{(*)} h_i$. However, this approach is not as trivial as it seems to be:

1. Each relay should spend at least M additional time slots to send the estimated g_i to the destination. Moreover, additional energy will be consumed when transmitting over additional time slots.
2. Transmitting the estimated g_i will suffer from further distortion because of both the noise at the destination and the error in the estimated channel h_i. Most time, g_i has to be quantized before the transmission [6], and the quantization error must also be counted.

Hence, a better way is to directly estimate the overall channel w_i at $\mathbb{D}$. We assume that the length of the training sequence sent from $\mathbb{S}$ is N, which may be different from the data block size T. The training sequence, denoted as $\mathbf{z}$, can be embedded into a data frame and will also be sent from $\mathbb{S}$ to $\mathbb{D}$ via the aid of $\mathbb{R}_i$'s. A linear transformation will be performed at each relay node before it forwards the training to the destination during Phase II. Denote the $N \times N$ unitary precoding matrix at the ith relay as $\mathbf{A}_i$ and define

$$\mathbf{C} = [\mathbf{A}_1 \mathbf{z}_1^{(*)}, \mathbf{A}_2 \mathbf{z}_2^{(*)}, \ldots, \mathbf{A}_M \mathbf{z}_M^{(*)}]. \qquad (2.23)$$

The transmitting model with other equations from (2.18) to (2.22) could be applied straightforwardly. With slight abuse of notations, we will keep all other notations unchanged from the previous section. During the training period, the power constraint is replaced by $\mathbf{z}^H \mathbf{z} \leq N P_s = E_s$.

1. *LS Estimation*: From (2.21), the LS estimate of $\mathbf{w}$ is derived as

$$\hat{\mathbf{w}}_{LS} = \boldsymbol{\Lambda}^{-1}\mathbf{C}^\dagger\mathbf{d}_2 = \mathbf{w} + \Delta\mathbf{w} \tag{2.24}$$

with error

$$\Delta\mathbf{w} = \boldsymbol{\Lambda}^{-1}\mathbf{C}^\dagger\mathbf{n}_d. \tag{2.25}$$

The covariance of $\Delta\mathbf{w}$ is then

$$\text{Cov}(\Delta\mathbf{w}|\mathbf{g}^{(*)}, \mathbf{h}) = \sigma_n^2\left(\sum_i |h_i|^2|\alpha_i|^2 + 1\right)\boldsymbol{\Lambda}^{-1}(\mathbf{C}^H\mathbf{C})^{-1}\boldsymbol{\Lambda}^{-1}, \tag{2.26}$$

where $\mathbf{g}^{(*)} = [g_1^{(*)}, g_2^{(*)}, \ldots, g_M^{(*)}]^T$ and $\mathbf{h} = [h_1, h_2, \ldots, h_M]^T$ are defined for convenience. Since $\boldsymbol{\Lambda}$ is a constant matrix, the optimization is conducted by varying the value of $\mathbf{C}$. Since the diagonal elements of $\mathbf{C}$ must all be no greater than E_s, the optimal $\mathbf{C}$ can be found by solving the following constrained optimization problem:

$$\min_{\boldsymbol{\Lambda}_i, \mathbf{z}} \quad \text{tr}\left(\boldsymbol{\Lambda}^{-1}(\mathbf{C}^H\mathbf{C})^{-1}\boldsymbol{\Lambda}^{-1}\right) \tag{2.27}$$
$$\text{s.t.} \quad [\mathbf{C}^H\mathbf{C}]_{ii} \leq E_s, \quad i = 1, \ldots M.$$

Note that, the above optimization problem is different from that of traditional multiple input single output (MISO) system, where there is a total power constraint over all transmit antennas [7]. In URS, since different relays could not share a common power pool, each relay will have its own power constraint P_{ri}, which is reflected by M individual constraints in (2.27).

2. *LMMSE Estimation*: Denote the covariance of $\mathbf{h}$ and $\mathbf{g}^{(*)}$ as $\mathbf{R}_h$ and $\mathbf{R}_{g(*)}$ respectively. Then, the covariance matrix of $\mathbf{w}$, assuming channels of Phase I are independent from channels of Phase II, is

$$\mathbf{R}_w = \text{E}\{\mathbf{w}\mathbf{w}^H\} = \mathbf{R}_{g(*)} \odot \mathbf{R}_h, \tag{2.28}$$

where $\odot$ denotes the Hadamard product. The LMMSE estimator of $\mathbf{w}$ is expressed as

$$\hat{\mathbf{w}}_{MMSE} = \text{E}\{\mathbf{w}\mathbf{d}_2^H\}(\text{E}\{\mathbf{d}_2\mathbf{d}_2^H\})^{-1}\mathbf{d}_2$$
$$= \mathbf{R}_w\boldsymbol{\Lambda}\mathbf{C}^H\left(\mathbf{C}\boldsymbol{\Lambda}\mathbf{R}_w\boldsymbol{\Lambda}\mathbf{C}^H + \sigma_n^2\sum_i(\sigma_{hi}^2|\alpha_i|^2 + 1)\mathbf{I}\right)^{-1}\mathbf{d}_2, \tag{2.29}$$

and the error covariance could also be obtained as

$$\text{Cov}(\Delta\mathbf{w}) = \left(\mathbf{R}_w^{-1} + \frac{1}{\sigma_n^2\sum_i(\sigma_{hi}^2|\alpha_i|^2 + 1)}\boldsymbol{\Lambda}\mathbf{C}^H\mathbf{C}\boldsymbol{\Lambda}\right)^{-1}. \tag{2.30}$$

The optimal training should then be obtained from

$$\min_{\mathbf{A}_i, \mathbf{z}} \quad \mathrm{tr}(\mathrm{Cov}(\Delta \mathbf{w})) \tag{2.31}$$

$$\text{s.t.} \quad [\mathbf{C}^H \mathbf{C}]_{ii} \leq E_s, \ i = 1, \ldots, M.$$

Remark. It is not difficult to see that the channel estimation as well as the corresponding training design in URS is quite different from the conventional point-to-point system since the insertion of relay node changes the whole transmitting configurations. One should try to recognize the "needed" channel for different schemes such that the estimation can be simplified and the resources consumed could be saved.

2.3   Challenges of Channel Estimation for PLNC

We have discussed the channel estimation in conventional point-to-point system as well as in URS in previous sections. Though the main difference between PLNC and URS is its bi-directional transmission, the channel estimation may, still, demonstrates much difference.

A typical PLNC model is presented in Fig. 2.2. We see that the channels from $\mathbb{T}_1$ to $\mathbb{R}$, denoted as $\mathbf{h}_1$, and that from $\mathbb{T}_2$ to $\mathbb{R}$, denoted as $\mathbf{h}_2$ can be estimated at $\mathbb{R}$ during the first phase, thanks to the simultaneous transmission from the two terminals. Most PLNC works focus on the TDD system that could save half of the bandwidth. Due to the reciprocity, the reverse channels from $\mathbb{R}$ to $\mathbb{T}_1$ and that from $\mathbb{R}$ to $\mathbb{T}_2$ remain $\mathbf{h}_1$ and $\mathbf{h}_2$, respectively. Based on these channel information, the relay node $\mathbb{R}$ could take on certain signal processing approach to optimize the overall performance of PLNC. Typical operations at $\mathbb{R}$ include

1. Beamforming design and power allocation [8];
2. Carrier permutation in an OFDM modulation [9].

Not like in URS, the task of channel estimation in PLNC should be obtaining the individual channels $\mathbf{h}_1$ and $\mathbf{h}_2$ at both $\mathbb{T}_1$ and $\mathbb{T}_2$ because of the following reasons:

Typically, the optimal operation at $\mathbb{R}$ varies according to the instant $\mathbf{h}_1$ and $\mathbf{h}_2$, while $\mathbb{T}_1$ and $\mathbb{T}_2$ must know the current signal operation at $\mathbb{R}$ in order to construct the overall "needed" channels. Hence, knowing $\mathbf{h}_1$ and $\mathbf{h}_2$ at $\mathbb{T}_1$ and $\mathbb{T}_2$ can help them predict the relay's operation and thus eliminates the necessity of the feedback channel.

Fig. 2.2 An example of general PLNC system

The most trivial way to obtain $\mathbf{h}_1$ and $\mathbf{h}_2$ at $\mathbb{T}_1$ and $\mathbb{T}_2$ is to ask $\mathbb{R}$ to send separate training sequence. As the training sequence is usually embedded in the data frame, one may hesitate to apply this way since it is not compatible with the two-phase transmission structure. Hence, the key challenge of the channel estimation in PLNC is **how to achieve the individual channel knowledge of $\mathbf{h}_1$ and $\mathbf{h}_2$ within two-phase training**.

In the next three chapters, we will present novel channel estimation schemes as well as their corresponding training design for PLNC under three typical scenarios: frequency flat fading environment, frequency selective environment, and time selective environment. We will see how the channel estimation differs in PLNC from that in the conventional point-to-point system or even from the unidirectional relay system.

References

1. C. S. Patel and G. L. Stuber. Channel estimation for amplify and forward relay based cooperation diversity systems. *IEEE Trans. on Wireless Commun.* 6(6), pp. 2348–2356, Aug. 2007.
2. Y. Jing, and B. Hassibi. Distributed space time coding in wireless relay networks. *IEEE Trans. Wireless Commun.*, 5(12), pp. 3524–3536, Dec. 2006.
3. Y. Jing, and H. Jafarkhani. Using orthogonal and quasi-orthogonal designs in wireless relay networks. in *Proc, GLOBECOM'06*, Nov. 2006.
4. J. N. Laneman, D. N. C. Tse, and G. W. Wornell. Cooperative diversity in wireless networks: efficient protocols and outage behavior. *IEEE Trans. Inform. Theory*, 50(12), pp. 3062–3080, Dec. 2004.
5. J. N. Laneman and G. W. Wornell. Distributed space time block coded protocols for exploiting cooperative diversity in wireless networks. *IEEE Trans. Inform. Theory*, 49(10), pp. 2415–2425, Oct. 2003.
6. J.-J. Xiao and Z.-Q. Luo. Universal decentralized estimation in an inhomogeneous sensing environment. *IEEE Trans. Inform. Theorey*, 51(10), pp. 3564–3575, Oct. 2005.
7. M. Biguesh, and A. B. Gershman. Training based MIMO channel estimation: a study of estimator tradeoffs and optimal training signals. *IEEE Trans. Signal Processing*, 54(3), pp. 884–893, Mar. 2006.
8. R. Zhang, Y.-C. Liang, C. C. Chai, and S. G. Cui. Optimal beamforming for two-way multi-antenna relay channel with analogue network coding. *IEEE J. Select. Areas in Commun.*, 27(5),pp. 699–712 , June 2009.
9. C. K. Ho, R. Zhang, and Y.-C. Liang. Two-way relaying over OFDM: optimized tone permutation and power allocation. in *Proc. of IEEE ICC*, pp. 3908–3912, Beijing China, May 2008.

Chapter 3
Channel Estimation for PLNC Under Frequency Flat Fading Scenario

Abstract In this chapter, we consider channel estimation for PLNC system in a frequency flat fading scenario. We propose a two-phase training protocol for channel estimation that can be easily embedded into the two-phase data transmission. Each terminal targets at estimating the individual channel parameters. We first derive the maximum-likelihood (ML) estimator, which is nonlinear and differs much from the conventional least-square (LS) estimator. Due to the difficulty in obtaining a closed-form expression of the mean square error (MSE) for the ML estimator, we resort to the Cramér-Rao lower bound (CRLB) of the estimation MSE to design the optimal training sequence. In the mean time, we introduce a new type of estimator that aims at maximizing the effective receive signal-to-noise ratio (SNR) after taking into consideration the channel estimation errors, referred to as the linear maximum signal-to-noise ratio (LMSNR) estimator. Furthermore, we prove that orthogonal training design is optimal for both the CRLB- and the LMSNR-based design criteria. Finally, simulations are presented to corroborate the proposed studies.

3.1 System Model

A typical transmission model of PLNC system with two terminals $\mathbb{T}_1$, $\mathbb{T}_2$, and one relay $\mathbb{R}$ is shown in Fig. 3.1. The baseband channel between $\mathbb{T}_i$ and $\mathbb{R}$ is denoted by h_i that is assumed as zero-mean circularly symmetric complex Gaussian random variable with variance $\sigma_{h_i}^2$. The average transmission powers of $\mathbb{T}_1$, $\mathbb{T}_2$, and $\mathbb{R}$ are P_1, P_2, and P_r, respectively.

Assume that $2N$ symbols are assigned for training, which could possibly be embedded in the data frame. Denote the training vector from $\mathbb{T}_1$ as $\mathbf{t}_1$ and the one from $\mathbb{T}_2$ as $\mathbf{t}_2$. Then, the received signal at $\mathbb{T}_1$ is[1]

$$\mathbf{z}_1 = \alpha h_1^2 \mathbf{t}_1 + \alpha h_1 h_2 \mathbf{t}_2 + \alpha h_1 \mathbf{n}_r + \mathbf{n}_1, \tag{3.1}$$

---

[1]Due to symmetry, we only present the channel estimation at $\mathbb{T}_1$.

© The Author(s) 2014

F. Gao et al., *Channel Estimation for Physical Layer Network Coding Systems*,
SpringerBriefs in Computer Science, DOI 10.1007/978-3-319-11668-6_3

Fig. 3.1 A typical PLNC
system with two terminals
and one relay node

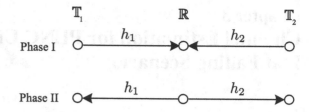

where $\mathbf{n}_r$ and $\mathbf{n}_1$ are the corresponding $N \times 1$ noise vectors at $\mathbb{R}$ and $\mathbb{T}_1$, respectively. For simplicity, the noise variances at all nodes are assumed as σ_n^2. Moreover, the scaling factor at $\mathbb{R}$ is set as

$$\alpha = \sqrt{\frac{P_r}{\sigma_{h_1}^2 P_1 + \sigma_{h_2}^2 P_2 + \sigma_n^2}} \tag{3.2}$$

to keep the relay power as P_r from a long term observation.

3.2   Maximum Likelihood Estimation

The PDF of the observed $\mathbf{z}_1$ is then

$$p(\mathbf{z}_1|h_1, h_2) = \frac{1}{\pi^N \sigma_n^{2N} (\alpha^2 |h_1|^2 + 1)^N} \exp\left(-\frac{\|\mathbf{z}_1 - \alpha h_1^2 \mathbf{t}_1 - \alpha h_1 h_2 \mathbf{t}_2\|^2}{\sigma_n^2 (\alpha^2 |h_1|^2 + 1)}\right), \tag{3.3}$$

and the corresponding likelihood function is

$$\log p(\mathbf{z}_1|h_1, h_2) = -\frac{\|\mathbf{z}_1 - \alpha h_1^2 \mathbf{t}_1 - \alpha h_1 h_2 \mathbf{t}_2\|^2}{\sigma_n^2 (\alpha^2 |h_1|^2 + 1)} - N \log(\alpha^2 |h_1|^2 + 1) - N \log(\pi \sigma_n^2). \tag{3.4}$$

Obviously, a direct ML estimation of h_1 and h_2 could be non-linear and is hard to implement.

3.2.1   Channel Estimation

Let us define new variables $a \triangleq h_1^2$, $b \triangleq h_1 h_2$ as the two equivalent channels, from which h_1 and h_2 could be recovered. The ML estimates of a, b can be obtained from

$$\{\hat{a}, \hat{b}\} = \arg\min_{a,b} \frac{\|\mathbf{z}_1 - \alpha a \mathbf{t}_1 - \alpha b \mathbf{t}_2\|^2}{\sigma_n^2 (\alpha^2 |a| + 1)} + N \log(\alpha^2 |a| + 1). \tag{3.5}$$

For comparison, the popular LS estimator is also displayed here:

$$\{\hat{a}, \hat{b}\} = \arg \min_{a,b} \|\mathbf{z}_1 - \alpha a \mathbf{t}_1 - \alpha b \mathbf{t}_2\|^2. \tag{3.6}$$

It is well known that LS approach in traditional point-to-point systems [1] coincides with the ML approach. However for PLNC, the LS approach never provides the ML estimation.

By observing (3.5), we know the ML estimate of b for given a can be obtained from

$$\hat{b} = \arg \min_{b} \|\mathbf{z}_1 - \alpha a \mathbf{t}_1 - \alpha b \mathbf{t}_2\|^2 = \frac{\mathbf{t}_2^H}{\alpha \|\mathbf{t}_2\|^2} (\mathbf{z}_1 - \alpha a \mathbf{t}_1). \tag{3.7}$$

Substituting (3.7) back to (3.5), we obtain

$$
\begin{aligned}
\hat{a} &= \arg \min_{a} \frac{\left\| \left(\mathbf{I} - \frac{\mathbf{t}_2 \mathbf{t}_2^H}{\|\mathbf{t}_2\|^2} \right) (\mathbf{z}_1 - \alpha a \mathbf{t}_1) \right\|^2}{\sigma_n^2 (\alpha^2 |a| + 1)} + N \log(\alpha^2 |a| + 1) \\
&= \arg \min_{a} \frac{\mathbf{z}_1^H \mathbf{A} \mathbf{z}_1 - 2\alpha \Re\{a \mathbf{z}_1^H \mathbf{A} \mathbf{t}_1\} + \alpha^2 |a|^2 \mathbf{t}_1^H \mathbf{A} \mathbf{t}_1}{\sigma_n^2 (\alpha^2 |a| + 1)} + N \log(\alpha^2 |a| + 1) \\
&= \arg \min_{a} \frac{\mathbf{z}_1^H \mathbf{A} \mathbf{z}_1 - 2\alpha |a| \Re\{e^{j \angle a} \mathbf{z}_1^H \mathbf{A} \mathbf{t}_1\} + \alpha^2 |a|^2 \mathbf{t}_1^H \mathbf{A} \mathbf{t}_1}{\sigma_n^2 (\alpha^2 |a| + 1)} + N \log(\alpha^2 |a| + 1),
\end{aligned}
\tag{3.8}
$$

where $\mathbf{A} \triangleq \mathbf{I} - \frac{\mathbf{t}_2 \mathbf{t}_2^H}{\|\mathbf{t}_2\|^2}$ is a projection matrix and $\angle a$ is the phase of a.

Note that $\angle a$ can be first estimated as

$$\widehat{\angle a} = -\angle(\mathbf{z}_1^H \mathbf{A} \mathbf{t}_1), \tag{3.9}$$

and then $|a|$ is obtained from

$$\widehat{|a|} = \arg \min_{x} \underbrace{\frac{\mathbf{z}_1^H \mathbf{A} \mathbf{z}_1 - 2\alpha x |\mathbf{z}_1^H \mathbf{A} \mathbf{t}_1| + \alpha^2 x^2 \mathbf{t}_1^H \mathbf{A} \mathbf{t}_1}{\sigma_n^2 (\alpha^2 x + 1)} + N \log(\alpha^2 x + 1)}_{f(x)} \tag{3.10}$$

$$\text{s.t.} \quad x \geq 0,$$

where $f(x)$ is defined as the corresponding objective function. After some tedious calculation, the derivative of $f(x)$ is computed as

$$\dot{f}(x) = \frac{\partial f(x)}{\partial x} = \frac{C_1 x^2 + C_2 x + C_3}{(\alpha^2 x + 1)^2}, \tag{3.11}$$

with

$$C_1 = \frac{\alpha^4 \mathbf{t}_1^H \mathbf{A} \mathbf{t}_1}{\sigma_n^2} = \frac{\alpha^4 (1 - |\rho|^2) \|\mathbf{t}_1\|^2}{\sigma_n^2},$$

$$C_2 = \alpha^4 N + \frac{2\alpha^2 \mathbf{t}_1^H \mathbf{A} \mathbf{t}_1}{\sigma_n^2} = \alpha^4 N + \frac{2\alpha^2 (1 - |\rho|^2) \|\mathbf{t}_1\|^2}{\sigma_n^2},$$

$$C_3 = \alpha^2 N - \frac{\alpha^2 \mathbf{z}_1^H \mathbf{A} \mathbf{z}_1 + 2\alpha |\mathbf{z}_1^H \mathbf{A} \mathbf{t}_1|}{\sigma_n^2},$$

and $\rho \triangleq \frac{\mathbf{t}_1^H \mathbf{t}_2}{\|\mathbf{t}_1\| \|\mathbf{t}_2\|}$ is the correlation factor between $\mathbf{t}_1$ and $\mathbf{t}_2$.

Case 1 ($|\rho| = 1$): When $\mathbf{t}_1$ and $\mathbf{t}_2$ are fully correlated, then $C_1 = 0$ and the solution to $\dot{f}(x) = 0$ is $x_0 = -\frac{C_3}{C_2}$. It can be checked that $\lim_{x \to +\infty} \dot{f}(x) > 0$ and $\lim_{x \to -\infty} \dot{f}(x) < 0$. So x_0 is the global minimal of $f(x)$. Considering that $|a| \geq 0$, the estimate of $|a|$ is

$$\widehat{|a|} = \max\{-C_3/C_2, 0\}. \tag{3.12}$$

However, fully correlated training should never be chosen since the summation of the training will become $\alpha(a + \sqrt{P_2/P_1} b e^{j\phi})\mathbf{t}_1$, where ϕ is the phase of ρ. In this case the two channels a, b can never be discriminated.

Case 2 ($|\rho| < 1$ and $C_2^2 - 4C_1 C_3 \geq 0$): When $\mathbf{t}_1$ and $\mathbf{t}_2$ are partially correlated or uncorrelated, $\dot{f}(x)$ has two roots which are derived as $x_{1,2} = \frac{-C_2 \pm \sqrt{C_2^2 - 4C_1 C_3}}{2C_1}$. It can be readily checked that $\lim_{x \to +\infty} \dot{f}(x) > 0$ and $\lim_{x \to -\infty} \dot{f}(x) > 0$. Therefore, $x_1 = \frac{-C_2 - \sqrt{C_2^2 - 4C_1 C_3}}{2C_1} < 0$ must be the local maximal and $x_2 = \frac{-C_2 + \sqrt{C_2^2 - 4C_1 C_3}}{2C_1}$ must be the local minimal. Considering that $|a| \geq 0$, the estimate of $|a|$ is

$$\widehat{|a|} = \max \left\{ \frac{-C_2 + \sqrt{C_2^2 - 4C_1 C_3}}{2C_1}, 0 \right\}. \tag{3.13}$$

Case 3 ($|\rho| < 1$ and $C_2^2 - 4C_1 C_3 < 0$): In this case, the derivative $\dot{f}(x)$ does not have roots. Then, $f(x)$ is a linearly increasing function whose minimum is achieved when $\widehat{|a|} = 0$.

Remark. When $C_3 \geq 0$, then $\hat{a}$ is 0. In this case, b can be estimated as $\hat{b} = \frac{\mathbf{t}_2^H}{\alpha \|\mathbf{t}_2\|^2} \mathbf{z}_1$, which coincides with the ML channel estimation in uni-directional relay system [2], as if $\mathbb{T}_1$ does not send out the training signal, although $\mathbb{T}_1$ does send out training here. This phenomenon is quite interesting since $\mathbb{T}_1$ does not see the channel h_1 but can still see the channel $h_1 h_2$. We call this as *hiding relay* scenario. Note that whether C_3 is greater than zero or not is non-predictable since it is determined by the instant value of the noise and the unknown channels. However, one can still gain some insight by considering the instant noise as 0, under which $C_3 \geq 0$ happens when

$$\frac{\|\mathbf{t}_1\|^2}{N\sigma_n^2} \leq \frac{1}{(1 - |\rho|^2)(\alpha^2|a| + 2)|a|}, \tag{3.14}$$

whose LHS can be considered as the average transmit SNR from $\mathbb{T}_1$ during the training. Therefore, many factors may result in a zero estimate of a, e.g., the deep fading between $\mathbb{T}_1$ and $\mathbb{R}$, the low training power from $\mathbb{T}_1$, the high correlation between $\mathbf{t}_1$ and $\mathbf{t}_2$, etc. Note that the relay scaling factor seems not that important to cause the *hiding relay* phenomenon because of the constant 2 in the denominator.

3.2.2 Training Sequence Design

Due to the nonlinearity in ML channel estimation, it is hard to represent the estimation MSE in closed form. As a result, we cannot design training sequence by minimizing MSE. Instead, we resort to CRLB (e.g., [3]) as our optimization criterion, which is another popular approach for training design and, most time, provides sufficient insight on the structure of the optimal training.

Theorem 3.1. *The CRLBs for a, b in PLNC are*

$$\text{CRLB}_a = \frac{A_3(A_1 A_3 - |A_2|^2)}{|A_2|^4 - 2A_1 A_3 |A_2|^2 + A_1^2 A_3^2 - |A_4|^2 A_3^2}, \tag{3.15}$$

$$\text{CRLB}_b = \frac{(-A_1|A_2|^2 - |A_4|^2 A_3 + A_1^2 A_3)}{|A_2|^4 - 2A_1 A_3 |A_2|^2 + A_1^2 A_3^2 - |A_4|^2 A_3^2}, \tag{3.16}$$

respectively, where

$$A_1 = \frac{\alpha^2 \|\mathbf{t}_1\|^2}{\sigma_n^2(\alpha^2|a| + 1)} + \frac{\alpha^4(N^2 - 2N + 2)}{4(\alpha^2|a| + 1)^2}, \qquad A_2 = \frac{\alpha^2 \rho \|\mathbf{t}_1\| \|\mathbf{t}_2\|}{\sigma_n^2(\alpha^2|a| + 1)},$$

$$A_3 = \frac{\alpha^2 \|\mathbf{t}_2\|^2}{\sigma_n^2(\alpha^2|a| + 1)}, \qquad A_4 = \frac{\alpha^4(N^2 - 2N + 2)a^2}{4(\alpha^2|a| + 1)^2|a|^2}.$$

Proof. See Appendix 1.   □

The derivatives of CRLB_a, CRLB_b with respective to $|A_2|^2$ are computed as

$$\frac{\partial \text{CRLB}_a}{\partial |A_2|^2} = \frac{A_3((|A_2|^2 - A_1 A_3)^2 + |A_4|^2 A_3^2)}{(|A_2|^4 - 2|A_2|^2 A_1 A_3 + A_1^2 A_3^2 - |A_4|^2 A_3^2)^2} > 0, \tag{3.17}$$

$$\frac{\partial \text{CRLB}_b}{\partial |A_2|^2} = \frac{A_1(|A_2|^2 - A_1 A_3)^2 + 2A_3|A_4|^2(|A_2|^2 - A_1 A_3) + A_1 A_3^2 |A_4|^2}{(|A_2|^4 - 2|A_2|^2 A_1 A_3 + A_1^2 A_3^2 - |A_4|^2 A_3^2)^2} > 0,$$

$$(3.18)$$

where the inequality in (3.18) is due to the fact that $|A_4|^2 < A_1^2$ so that the discriminant of the numerator is smaller than zero. Therefore, both CRLB_a and CRLB_b are increasing functions of $|\rho|^2$ and reach their minimum when $|\rho|^2 = 0$. Then the optimal $\mathbf{t}_1$, $\mathbf{t}_2$ should be orthogonal, and the corresponding CRLBs are given by

$$\text{CRLB}_a = \frac{A_1}{A_1^2 - |A_4|^2}, \tag{3.19}$$

$$\text{CRLB}_b = \frac{1}{A_3}. \tag{3.20}$$

Suppose the maximum power assigned to $\mathbf{t}_i$ is Q_i (typically NP_i). The optimal training power to minimize CRLB_b is obviously Q_2; namely, $\mathbb{T}_2$ should always transmit at its maximum power. The optimization for $\|\mathbf{t}_1\|^2$ becomes

$$\|\mathbf{t}_1\|^2 = \arg\min_x \frac{x + c}{x^2 + 2cx} \tag{3.21}$$

$$\text{s.t.} \quad 0 \leq x \leq Q_1,$$

where $c = \frac{\alpha^2 \sigma_n^2 (N^2 - 2N + 2)}{4(\alpha^2 |a| + 1)}$. It can be easily shown that the objective is a linear decreasing function of x, so the optimal training power for $\mathbb{T}_1$ is the maximum power Q_1.

Remark. Note that LS estimator is an unbiased estimator whose error covariance matrix can be derived as

$$\text{Cov}_{\text{LS}} = \frac{\sigma_n^2(\alpha^2 |a| + 1)}{\alpha^2} \begin{bmatrix} \|\mathbf{t}_1\|^2 & \rho \|\mathbf{t}_1\| \|\mathbf{t}_2\| \\ \rho^* \|\mathbf{t}_2\| \|\mathbf{t}_1\| & \|\mathbf{t}_2\|^2 \end{bmatrix}^{-1}. \tag{3.22}$$

Denote the error variance of a, b from LS as cov_a and cov_b, respectively. It is not hard to know that if $\rho > 0$, then $\text{CRLB}_a < \text{cov}_a$ and $\text{CRLB}_b < \text{cov}_b$. Therefore, LS is never an optimal choice when $\rho > 0$ for PLNC. However, if $\rho = 0$, then $\text{CRLB}_b = \text{cov}_b$ but $\text{CRLB}_a < \text{cov}_a$. In this case, the LS estimation is optimal in terms of estimating b but is still non-optimal in terms of estimating a.

3.3 Linear Maximum Signal-to-Noise Ratio Based Estimation

Another popular channel estimation method is the LMMSE estimator that targets at minimizing the estimation MSE. However, this criterion may not be the ultimately good one since the final target of the communication is the data detection. Hence, we design a new linear channel estimator to maximize the effective SNR, which takes the channel estimation errors into consideration.

3.3.1 Channel Estimation

First note the following properties of a and b, where the fact that h_1 is independent of h_2 is used:

$$\mathrm{E}\{a\} = \mathrm{E}\{b\} = 0, \qquad \sigma_a^2 = \mathrm{E}\{|a|^2\} = 2\sigma_{h_1}^4, \qquad \sigma_b^2 = \mathrm{E}\{|b|^2\} = \sigma_{h_2}^2 \sigma_{h_1}^2,$$

$$\mathrm{E}\{ab^*\} = 0, \qquad \mathrm{E}\{ah_1^*\} = 0, \qquad \mathrm{E}\{bh_1^*\} = 0.$$

Suppose the linear estimations of a, b are $\hat{a} = \mathbf{u}^H \mathbf{z}_1$ and $\hat{b} = \mathbf{v}^H \mathbf{z}_1$, where $\mathbf{u}$ and $\mathbf{v}$ are the unknown vectors to be designed. Define $\Delta_a = \hat{a} - a$ and $\Delta_b = \hat{b} - b$.

During the data transmission, suppose $\mathbb{T}_i$ sends d_i to the other terminal. At $\mathbb{T}_1$, the remaining signal after canceling d_1 is

$$(y_1 - \alpha \hat{a} d_1) = \alpha \hat{b} d_2 + \tilde{n}, \tag{3.23}$$

where $\tilde{n} = -\alpha \Delta_b d_2 - \alpha \Delta_a d_1 + \alpha h n_r + n_1$ is the equivalent noise term that includes the channel estimation errors. The instant effective SNR and the average effective SNR (AESNR) of the detection are then expressed as

$$\gamma = \frac{\alpha^2 |\hat{b}|^2 P_2}{\mathrm{E}\{|\tilde{n}|^2\}} = \frac{|\mathbf{v}^H \mathbf{z}_1|^2 P_2}{\mathrm{E}\{|\mathbf{v}^H \mathbf{z}_1 - b|^2\} P_2 + \mathrm{E}\{|\mathbf{u}^H \mathbf{z}_1 - a|^2\} P_1 + \sigma_n^2 (\sigma_{h_1}^2 + 1/\alpha^2)}, \tag{3.24}$$

$$\bar{\gamma} = \frac{\mathrm{E}\{|\mathbf{v}^H \mathbf{z}_1|^2\} P_2}{\mathrm{E}\{|\mathbf{v}^H \mathbf{z}_1 - b|^2\} P_2 + \mathrm{E}\{|\mathbf{u}^H \mathbf{z}_1 - a|^2\} P_1 + \sigma_n^2 (\sigma_{h_1}^2 + 1/\alpha^2)}, \tag{3.25}$$

respectively. Note that the symbol error rate (SER) is closely related with AESNR as shown in [4].

Observing from $\bar{\gamma}$ that $E\{|\mathbf{u}^H\mathbf{z}_1 - a|^2\}$ is an independent factor, the optimal $\mathbf{u}$ is directly obtained from the LMMSE approach[2] as

$$\mathbf{u} = E^{-1}\{\mathbf{z}_1\mathbf{z}_1^H\}E\{a^*\mathbf{z}_1\} = \alpha\sigma_a^2\mathbf{R}_z^{-1}\mathbf{t}_1, \qquad (3.26)$$

where $\mathbf{R}_z = \alpha^2\sigma_a^2\mathbf{t}_1\mathbf{t}_1^H + \alpha^2\sigma_b^2\mathbf{t}_2\mathbf{t}_2^H + (\alpha^2\sigma_{h_1}^2 + 1)\sigma_n^2\mathbf{I}$ is the covariance matrix of $\mathbf{z}_1$. The MSE of a, defined as $e_a = E\{|\Delta_a|^2\}$, can be derived as

$$e_a = \sigma_a^2 - \alpha^2\sigma_a^4\mathbf{t}_1^H\mathbf{R}_z^{-1}\mathbf{t}_1 = \frac{\frac{\|\mathbf{t}_2\|^2}{\alpha^2} + \frac{\beta}{\alpha^4\sigma_b^2}}{\frac{\|\mathbf{t}_1\|^2\|\mathbf{t}_2\|^2(1-|\rho|^2)}{\beta} + \frac{\|\mathbf{t}_1\|^2}{\alpha^2\sigma_b^2} + \frac{\|\mathbf{t}_2\|^2}{\alpha^2\sigma_a^2} + \frac{\beta}{\alpha^4\sigma_a^2\sigma_b^2}}, \qquad (3.27)$$

where $\beta \triangleq (\alpha^2\sigma_{h_1}^2 + 1)\sigma_n^2$. Then the AESNR becomes

$$\bar{\gamma} = \frac{\mathbf{v}^H\mathbf{R}_z\mathbf{v}}{\mathbf{v}^H\mathbf{R}_z\mathbf{v} - B_1(\mathbf{v}^H\mathbf{t}_2 + \mathbf{t}_2^H\mathbf{v}) + B_2}, \qquad (3.28)$$

where

$$B_1 = \alpha\sigma_b^2, \qquad B_2 = \sigma_b^2 + e_a P_1/P_2 + \beta/(\alpha^2 P_2).$$

Suppose the maximum $\bar{\gamma}$ appears at $\mathbf{v} = \mathbf{v}_0$ and define $C_0 = \mathbf{v}_0^H\mathbf{R}_z\mathbf{v}_0$. Among all those $\mathbf{v}$'s that give $\mathbf{v}^H\mathbf{R}_z\mathbf{v} = C_0$, $\mathbf{v}_0$ must yield the smallest denominator value. Hence, we can first assign an arbitrary value C to $\mathbf{v}^H\mathbf{R}_z\mathbf{v}$ and then find the optimal $\mathbf{v}_C$ that minimizes the denominator. After testing all $C \geq 0$, the one gives the largest $\bar{\gamma}$ is chosen as C_0 and the corresponding $\mathbf{v}_C$ is the optimal design of $\mathbf{v}$.

Proposition 3.1. *The optimization*

$$\max_{\mathbf{v}} \quad \frac{\mathbf{v}^H\mathbf{R}_z\mathbf{v}}{\mathbf{v}^H\mathbf{R}_z\mathbf{v} - B_1(\mathbf{v}^H\mathbf{t}_2 + \mathbf{t}_2^H\mathbf{v}) + B_2} \qquad (3.29)$$

$$\text{s.t.} \quad \mathbf{v}^H\mathbf{R}_z\mathbf{v} = C,$$

is equivalent to

$$\min_{\mathbf{v}} \quad -\mathbf{v}^H\mathbf{t}_2 - \mathbf{t}_2^H\mathbf{v} \qquad (3.30)$$

$$\text{s.t.} \quad \mathbf{v}^H\mathbf{R}_z\mathbf{v} \leq C.$$

---

[2]Similarly, the LMMSE estimator of $\mathbf{v}$ is $\mathbf{v} = \alpha\sigma_b^2\mathbf{R}_z^{-1}\mathbf{t}_2$, which will be used in the later simulations.

Proof. It is easily known that the optimization in (3.29) is equivalent to

$$\min_{\mathbf{v}} \quad -\mathbf{v}^H \mathbf{t}_2 - \mathbf{t}_2^H \mathbf{v} \tag{3.31}$$

$$\text{s.t.} \quad \mathbf{v}^H \mathbf{R}_z \mathbf{v} = C.$$

Note that the optimal value of (3.31) must be smaller than zero. This is seen by choosing $\mathbf{v} = \sqrt{\frac{C}{\mathbf{t}_2^H \mathbf{R}_z \mathbf{t}_2}} \mathbf{t}_2$, which results in a negative value of the cost function. Hence, we can use the new constraint $\mathbf{v}^H \mathbf{R}_z \mathbf{v} \leq C$ instead, because at the optimal point the equality always holds; otherwise, we can scale the value of $\mathbf{v}$ by a positive factor to keep the equality while getting an even smaller value of cost function. □

The optimization (3.30) is convex and the solution can be found from the Lagrange dual function [5] as

$$\mathbf{v}_C = \frac{\sqrt{C} \mathbf{R}_z^{-1} \mathbf{t}_2}{\sqrt{\mathbf{t}_2^H \mathbf{R}_z^{-1} \mathbf{t}_2}}. \tag{3.32}$$

Substituting $\mathbf{v}_C$ back to (3.29) gives

$$\bar{\gamma} = \frac{C}{C - 2B_1 \sqrt{\mathbf{t}_2^H \mathbf{R}_z^{-1} \mathbf{t}_2} \sqrt{C} + B_2}. \tag{3.33}$$

Then, the optimal $\sqrt{C_0}$ can be calculated as

$$\sqrt{C_0} = \frac{B_2}{B_1 \sqrt{\mathbf{t}_2^H \mathbf{R}_z^{-1} \mathbf{t}_2}}. \tag{3.34}$$

Finally, the optimal linear estimator $\mathbf{v}$ is

$$\mathbf{v} = \frac{B_2 \mathbf{R}_z^{-1} \mathbf{t}_2}{B_1 \mathbf{t}_2^H \mathbf{R}_z^{-1} \mathbf{t}_2}. \tag{3.35}$$

3.3.2 Training Sequence Design

Substituting C_0 back to (3.29) gives the expression of maximum AESNR as

$$\bar{\gamma} = \frac{B_2}{B_2 - B_1^2 \mathbf{t}_2^H \mathbf{R}_z^{-1} \mathbf{t}_2}, \tag{3.36}$$

where the explicit form of $\mathbf{t}_2^H \mathbf{R}_z^{-1} \mathbf{t}_2$ can be calculated as

$$\mathbf{t}_2^H \mathbf{R}_z^{-1} \mathbf{t}_2 = \frac{\frac{(1-|\rho|^2)\|\mathbf{t}_1\|^2\|\mathbf{t}_2\|^2}{\alpha^2 \sigma_b^2 \beta} + \frac{\|\mathbf{t}_2\|^2}{\alpha^4 \sigma_a^2 \sigma_b^2}}{\frac{(1-|\rho|^2)\|\mathbf{t}_1\|^2\|\mathbf{t}_2\|^2}{\beta} + \frac{\|\mathbf{t}_1\|^2}{\alpha^2 \sigma_b^2} + \frac{\|\mathbf{t}_2\|^2}{\alpha^2 \sigma_a^2} + \frac{\beta}{\alpha^4 \sigma_a^2 \sigma_b^2}}. \tag{3.37}$$

After tedious re-organization, the AESNR becomes a function of $|\rho|^2$ as

$$\bar{\gamma} = \frac{D_1(1 - |\rho|^2) + D_2}{D_3(1 - |\rho|^2) + D_4}, \tag{3.38}$$

where

$$D_1 = \left(\frac{\sigma_b^2}{\beta} + \frac{1}{\alpha^2 P_2} \right) \|\mathbf{t}_1\|^2 \|\mathbf{t}_2\|^2,$$

$$D_2 = \frac{P_1}{P_2} \left(\frac{\|\mathbf{t}_2\|^2}{\alpha^2} + \frac{\beta}{\alpha^4 \sigma_b^2} \right) + \left(\sigma_b^2 + \frac{\beta}{\alpha^2 P_2} \right) \left(\frac{\|\mathbf{t}_1\|^2}{\alpha^2 \sigma_b^2} + \frac{\|\mathbf{t}_2\|^2}{\alpha^2 \sigma_a^2} + \frac{\beta}{\alpha^4 \sigma_a^2 \sigma_b^2} \right),$$

$$D_3 = \frac{\|\mathbf{t}_1\|^2 \|\mathbf{t}_2\|^2}{\alpha^2 P_2}, \qquad D_4 = D_2 - \frac{\sigma_b^2 \|\mathbf{t}_2\|^2}{\alpha^2 \sigma_a^2} \geq 0.$$

Note that $\bar{\gamma} \geq 1$ regardless of the values of $\|\mathbf{t}_1\|$, $\|\mathbf{t}_2\|$ and ρ.

It can be readily verified that $D_1, D_2, D_3, D_4 \geq 0$, and

$$D_2 D_3 - D_1 D_4$$

$$= - \|\mathbf{t}_1\|^2 \|\mathbf{t}_2\|^2 \left(\frac{\|\mathbf{t}_1\|^2}{\alpha^4 P_2} + \frac{P_1 \|\mathbf{t}_2\|^2 \sigma_b^2}{P_2 \alpha^2 \beta} + \frac{P_1}{P_2 \alpha^4} + \frac{\sigma_b^2 \|\mathbf{t}_1\|^2}{\alpha^2 \beta} + \frac{\sigma_b^2}{\alpha^4 \sigma_a^2} + \frac{\beta}{\alpha^6 P_2 \sigma_a^2} \right).$$

Therefore, the optimal $|\rho|$ should be zero since the function $\frac{x + D_2/D_1}{x + D_4/D_3}$ is maximized when $x = 1$ over $x \in [0, 1]$. The AESNR is then rewritten as

$$\bar{\gamma} = \frac{D_1 + D_2}{D_3 + D_4}. \tag{3.39}$$

Suppose the maximum training power of $\mathbb{T}_i$ is Q_i (typically NP_i). We need to find the optimal $\|\mathbf{t}_i\|^2 \in [0, Q_i]$ that maximize $\bar{\gamma}$. By noting that both the denominator and numerator in (3.39) are the function of $\|\mathbf{t}_1\|^2$, $\|\mathbf{t}_2\|^2$, and $\|\mathbf{t}_1\|^2 \|\mathbf{t}_2\|^2$, we can rewrite (3.39) as

$$\bar{\gamma}_M = F_0 \frac{\|\mathbf{t}_1\|^2 + \frac{F_1 \|\mathbf{t}_2\|^2}{F_2 \|\mathbf{t}_2\|^2 + F_3}}{\|\mathbf{t}_1\|^2 + \frac{F_4 \|\mathbf{t}_2\|^2}{F_5 \|\mathbf{t}_2\|^2 + F_6}}, \tag{3.40}$$

where F_i's are all positive coefficients whose explicit forms are omitted for brevity. Obviously, at the optimal point, $\|\mathbf{t}_1\|^2$ should be either 0 or Q_1 depending on the value of $\|\mathbf{t}_2\|^2$. Similarly, $\|\mathbf{t}_2\|^2$ should be either 0 or Q_2 depending on the value of $\|\mathbf{t}_1\|^2$. Therefore, the optimal solution may come from four possible sets $\{\|\mathbf{t}_1\|^2 = Q_1, \|\mathbf{t}_2\|^2 = Q_2\}$, $\{\|\mathbf{t}_1\|^2 = 0, \|\mathbf{t}_2\|^2 = Q_2\}$, $\{\|\mathbf{t}_1\|^2 = Q_1, \|\mathbf{t}_2\|^2 = 0\}$ and $\{\|\mathbf{t}_1\|^2 = 0, \|\mathbf{t}_2\|^2 = 0\}$. It can be checked that the third and the fourth sets always yield $\bar{\gamma} = 1$. Hence, only the first two sets need to be compared. From straightforward comparison, we know the first set gives a higher AESNR than the second. Therefore the optimal power allocation is simply to fully utilize the training power, as in the traditional point-to-point systems.

Remark. From our optimization over $\mathbb{T}_1$, we know that the training should be orthogonal and both the terminals should train at their maximum power. Similar optimization can be drawn for $\mathbb{T}_2$. Note that this point is quite important because if we meet contradicted power allocation when separately analyzing $\mathbb{T}_1$ and $\mathbb{T}_2$, then the overall optimal training has to be obtained by jointly considering the two terminals.

3.4 Numerical Results

In this section, we numerically study the performance of our proposed channel estimation algorithms and the training designs under various scenarios. The channels h_1, h_2 and the noise are assumed as circularly symmetric complex Gaussian random variables with unit variances. We set $P_2 = 2P_1$, $P_r = (P_1 + P_2)/2$, and the SNR is defined as P_2/σ_n^2. The parameter N is set as 8 and the phase of ρ is randomly taken. Totally 10^5 Monte-Carlo runs are adopted for average.

1. In the first example, we compare the performance of the ML channel estimation with that of the LS method. The channel estimation MSE is used as the figure of merit. The estimation results for a and b are separately shown in Figs. 3.2 and 3.3, and the CRLBs for different cases are displayed as well. It is seen that the ML estimation outperforms the LS estimation all the time. However, the difference between the two methods is reduced when $|\rho|$ becomes smaller or when SNR becomes higher. Specifically, when $\rho = 0$, the difference between ML method and LS method almost vanishes. As we have analyzed previously, the LS estimate of b can reach CRLB for $\rho = 0$.

   We also see that the ML estimator is a biased estimator in the two-way case since its MSE is lower than CRLB at some SNR region. The main reason is that we clip $\widehat{|a|}$ to zero for some realization of the noise, where we meet the *hiding relay* scenario. Therefore, $|\rho| = 0$ may not be the optimal correlation factor for the ML estimator. Nevertheless, the best $|\rho|$ for the channel estimation, from the numerical results, seems to be 0. Moreover, LS method is always lower bounded by CRLB since it is an unbiased estimator.

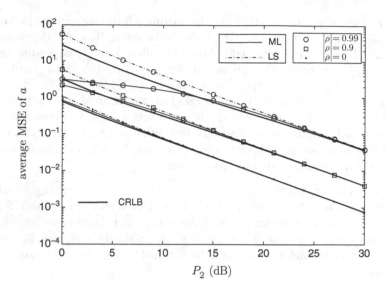

Fig. 3.2 Channel estimation MSEs versus SNR for a by ML and LS methods

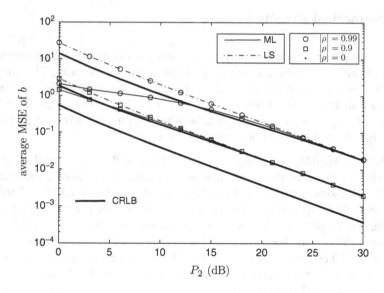

Fig. 3.3 Channel estimation MSEs versus SNR for b by ML and LS methods

2. We then examine the performance of the proposed LMSNR method in terms of the maximum AESNR that can be reached. The performance of AESNR from LMMSE channel estimation is also included for comparison. From Fig. 3.4, we see that the proposed LMSNR can provide higher AESNR than LMMSE. The SNR gain of LMSNR is larger at a relative lower SNR region for all different

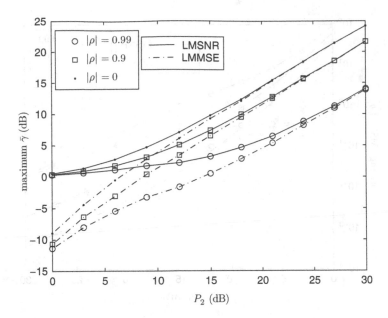

Fig. 3.4 Maximum average effective SNR versus SNR for LMSNR and LMMSE methods

values of $|\rho|$. At very high SNR region, the difference becomes smaller. As analyzed previously, AESNR from LMSNR method is always greater than or equal to 1, which does not hold for LMMSE estimator. It is also seen that the orthogonal training can provide the highest effective SNR, which validate our analytical study.

It is also interesting to take a look at the performance of the channel estimation for both LMSNR and LMMSE. In Fig. 3.5, we display the channel estimation MSE for b from the two methods. It is not surprising to see that LMMSE outperforms LMSNR in terms of the channel estimation MSE since it itself comes from the minimizing MSE criterion. However, even with such a big gain in channel estimation MSE, the LMMSE performs worse in terms of the AESNR.

3.5 Summary

In this chapter, we studied the channel estimation and the training design for PLNC under frequency flat fading environment. We first designed the ML estimation algorithm, which is seen to be different from the traditional LS channel estimator. Due to the nonlinearity of ML approach, we design the training sequence by resorting to CRLB. We further proposed a linear estimator called LMSNR, which aims at maximizing the AESNR. Interestingly, the orthogonal training with maximum

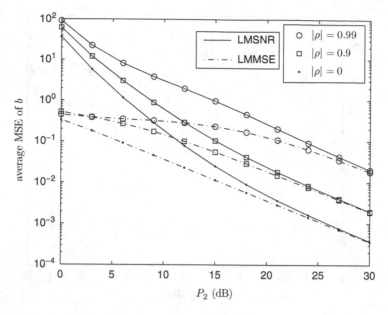

Fig. 3.5 Channel estimation MSE versus SNR for b by LMSNR and LMMSE methods

power transmission was proved to be optimal for both CRLB- and LMSNR-based design. Numerical examples clearly demonstrated the effectiveness of the proposed two estimators.

Appendix 1: Derivation of CRLB

Let $\boldsymbol{\theta} = [a, b]^T$ be the parameter vector to be estimated. In order to obtain the complex Fisher information matrix (FIM), we need to further define [6]

$$\mathbf{J}_{\varphi\psi} = \mathrm{E}\left\{\left(\frac{\partial \log p(\mathbf{z}_1|\boldsymbol{\theta})}{\partial \boldsymbol{\varphi}^*}\right)\left(\frac{\partial \log p(\mathbf{z}_1|\boldsymbol{\theta})}{\partial \boldsymbol{\psi}^*}\right)^H\right\}, \qquad (3.41)$$

where $\boldsymbol{\varphi}$ and $\boldsymbol{\psi}$ represent $\boldsymbol{\theta}$ or $\boldsymbol{\theta}^*$. Moreover, the derivative with respect to the complex vector $\boldsymbol{\theta} = \Re\{\boldsymbol{\theta}\} + j\Im\{\boldsymbol{\theta}\}$ is defined as $\frac{\partial}{\partial\boldsymbol{\theta}} = \frac{1}{2}\left(\frac{\partial}{\partial\Re\{\boldsymbol{\theta}\}} - j\frac{\partial}{\partial\Im\{\boldsymbol{\theta}\}}\right)$ and $\frac{\partial}{\partial\boldsymbol{\theta}^*} = \frac{1}{2}\left(\frac{\partial}{\partial\Re\{\boldsymbol{\theta}\}} + j\frac{\partial}{\partial\Im\{\boldsymbol{\theta}\}}\right)$.

Then, the FIM on $\boldsymbol{\theta}_R = [\Re\{\boldsymbol{\theta}\}^T, \Im\{\boldsymbol{\theta}\}^T]^T$ can be obtained as [7]

$$\mathbf{J}_{\boldsymbol{\theta}_R\boldsymbol{\theta}_R} = \mathcal{M}\begin{bmatrix}\mathbf{J}_{\theta\theta} & \mathbf{J}_{\theta\theta^*} \\ \mathbf{J}_{\theta\theta^*}^* & \mathbf{J}_{\theta\theta}^*\end{bmatrix}\mathcal{M}^H, \qquad (3.42)$$

where $\mathcal{M} = \begin{bmatrix} \mathbf{I} & \mathbf{I} \\ -j\mathbf{I} & j\mathbf{I} \end{bmatrix}$. The CRLB of θ_R is then

$$\text{CRLB} = \mathbf{J}_{\theta_R\theta_R}^{-1}. \tag{3.43}$$

By straightforward calculation, we obtain

$$\frac{\partial \log p(\mathbf{z}_1|a,b)}{\partial a^*} = \left(\frac{\partial \log p(\mathbf{z}_1|a,b)}{\partial a} \right)^*$$

$$= \frac{\alpha(\alpha h\bar{\mathbf{n}}_r + \bar{\mathbf{n}}_1)^H \mathbf{t}_1}{\sigma_n^2(\alpha^2|a|+1)} + \frac{|(\alpha h\bar{\mathbf{n}}_r + \bar{\mathbf{n}}_1)|^2 \alpha^2 a^*}{2\sigma_n^2(\alpha^2|a|+1)^2|a|} - \frac{N\alpha^2 a^*}{2(\alpha^2|a|+1)|a|},$$

$$\frac{\partial \log p(\mathbf{z}_1|a,b)}{\partial b^*} = \left(\frac{\partial \log p(\mathbf{z}_1|a,b)}{\partial b} \right)^* = \frac{\alpha(\alpha h\bar{\mathbf{n}}_r + \bar{\mathbf{n}}_1)^H \mathbf{t}_2}{\sigma_n^2(\alpha^2|a|+1)}.$$

Therefore,

$$[\mathbf{J}_{\theta\theta}]_{11} = A_1, \qquad [\mathbf{J}_{\theta\theta}]_{12} = [\mathbf{J}_{\theta\theta}]_{21}^* = A_2, \qquad\qquad [\mathbf{J}_{\theta\theta}]_{22} = A_3,$$

$$[\mathbf{J}_{\theta\theta^*}]_{11} = A_4, \qquad [\mathbf{J}_{\theta\theta^*}]_{12} = [\mathbf{J}_{\theta\theta^*}]_{21}^* = [\mathbf{J}_{\theta\theta^*}]_{22} = 0.$$

After tedious calculation, the CRLBs of a, b, defined as CRLB_a, CRLB_b, are

$$\text{CRLB}_a = [\text{CRLB}]_{11} + [\text{CRLB}]_{33}, \tag{3.44}$$

$$\text{CRLB}_b = [\text{CRLB}]_{22} + [\text{CRLB}]_{44}, \tag{3.45}$$

and the results are shown in (3.15) and (3.16), respectively.

References

1. H. Minn and N. Al-Dhahir. Optimal training signals for MIMO OFDM channel estimation. *IEEE Trans. Wireless Commun.*, 5(6), pp. 1158–1168, May 2006.
2. F. Gao, T. Cui, and A. Nallanathan. On channel estimation and optimal training design for amplify and forward relay network. *IEEE Trans. Wireless Commun.*, 7(5), pp. 1907–1916, May 2008.
3. P. Stoica and O. Besson. Training sequence design for frequency offset and frequency-selective channel estimation. *IEEE Trans. Commun.*, 51(11), pp. 1910–1917, Nov. 2003.
4. C. S. Patel and L. Stüber. Channel estimation for amplify and forward relay based cooperation diversity systems. *IEEE Trans. Wireless Commun.*, 6(6), pp. 2348–2356, June 2007.
5. S. Boyd and L. Vandenberghe, *Convex Optimization*, Cambridge University Press, 2006.
6. S. M. Kay, *Fundamentals of Statistical Signal Processing: Estimation Theory.* Englewood Cliffs, NJ: Prentice-Hall, 1993.
7. E. de Carvalho, J. Cioffi, and D. Slock. Cramér-rao bounds for blind multichannel estimation. in *Proc. of IEEE GLOBECOM*, pp. 1036–1040, San Francisco, Nov. 2000.

Chapter 4
Channel Estimation for PLNC Under Frequency Selective Fading Scenario

Abstract In this chapter, we consider the channel estimation for PLNC under the more general frequency selective scenario, where orthogonal-frequency-division multiplexing (OFDM) is adopted for data transmission. We propose a two-phase training protocol, which is compatible with the two-phase data transmission, and thus the training block can be embedded into the data frame. Specifically, we design two different types of training methods: (i) *block* based training, for which we first estimate the *cascaded* source-relay-source channels, and then recover the *individual* channels between sources and relay; (ii) *pilot-tone* (PT) based training, for which we directly estimate the individual channels between sources and relay. Importantly, the identifiability of the channel estimation in both types of the training schemes are fully addressed. Finally, various numerical examples are presented to corroborate our analytical results.

4.1 System Model

A typical PLNC under frequency selective environment is shown in Fig. 4.1. The baseband channel between $\mathbb{T}_1$ and $\mathbb{R}$ is denoted by $\mathbf{h} = [h_0, h_1, \ldots, h_{L_1-1}]^T$ and the one between $\mathbb{T}_2$ and $\mathbb{R}$ is denoted by $\mathbf{g} = [g_0, g_1, \ldots, g_{L_2-1}]^T$, where L_j represents the number of the taps of the corresponding channel.[1] The elements in $\mathbf{h}$ and $\mathbf{g}$ are assumed as zero-mean circularly symmetric complex Gaussian (CSCG) random vectors and are independent from one another. Particularly, the variance of the lth elements in $\mathbf{h}$ and $\mathbf{g}$ are denoted by $\sigma_{h,l}^2$ and $\sigma_{g,l}^2$, respectively. For simplicity, we consider that the taps within one channel vector are also independent from one another. Moreover, the time-division-duplexing (TDD) is a generally adopted assumption for PLNC [1–3], where the channels can be considered reciprocal such that the channel from $\mathbb{R}$ to $\mathbb{T}_1$ is still $\mathbf{h}$ and the channel from $\mathbb{R}$ to $\mathbb{T}_2$ is still $\mathbf{g}$. The average transmission powers of $\mathbb{T}_1$, $\mathbb{T}_2$, and $\mathbb{R}$ are denoted as P_1, P_2, and P_r, respectively.

---

[1]Note that we do not use $\mathbf{h}_1$ and $\mathbf{h}_2$ in this chapter to avoid the heavy notations.

© The Author(s) 2014

F. Gao et al., *Channel Estimation for Physical Layer Network Coding Systems*,
SpringerBriefs in Computer Science, DOI 10.1007/978-3-319-11668-6_4

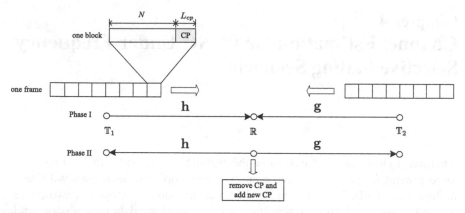

Fig. 4.1 A typical PLNC system adopting OFDM modulation

4.1.1 OFDM Transmission in PLNC

To combat the frequency-selective fading channels, we apply the well-known orthogonal frequency division multiplexing (OFDM) [4] for transmission. As shown in Fig. 4.1, each frame contains multiple OFDM blocks, while each OFDM block contains N information symbols and a cyclic prefix (CP) of length L_{cp}. To avoid the inter-block interference, L_{cp} should be greater than or equal to $\max\{L_1 - 1, L_2 - 1\}$. Denote the information vector of one OFDM block from $\mathbb{T}_j$ as $\tilde{\mathbf{d}}_j = [\tilde{d}_{j,0}, \tilde{d}_{j,1}, \ldots, \tilde{d}_{j,N-1}]^T$, $j = 1, 2$. The corresponding time-domain signal vector is obtained from the normalized inverse discrete Fourier transformation (IDFT) as

$$\mathbf{d}_j = \mathbf{F}^H \tilde{\mathbf{d}}_j = [d_{j,0}, d_{j,1}, \ldots, d_{j,N-1}]^T, \qquad (4.1)$$

where $\mathbf{F}$ is the matrix with the (p, q)th entry given by $\frac{1}{\sqrt{N}} e^{-j2\pi(p-1)(q-1)/N}$. After the CP insertion, $\mathbb{T}_j$'s send out the resultant signal simultaneously during Phase I. The received signal at $\mathbb{R}$, after removing CP, can be represented by

$$\mathbf{r} = [r_0, r_1, \ldots, r_{N-1}]^T = \mathbf{H}\mathbf{d}_1 + \mathbf{G}\mathbf{d}_2 + \mathbf{n}_r, \qquad (4.2)$$

where $\mathbf{n}_r = [n_{r,0}, n_{r,1}, \ldots, n_{r,N-1}]^T$ denotes the zero-mean additive-white-Gaussian-noise (AWGN), each element with variance σ_n^2. In addition, $\mathbf{H}$ and $\mathbf{G}$ are the $N \times N$ circulant matrices whose first columns are $[\mathbf{h}^T, \mathbf{0}_{1 \times (N-L_1)}]^T$ and $[\mathbf{g}^T, \mathbf{0}_{1 \times (N-L_2)}]^T$, respectively. The normalized discrete Fourier transformation (DFT) of $\mathbf{r}$ is denoted by

$$\tilde{\mathbf{r}} = [\tilde{r}_0, \tilde{r}_1, \ldots, \tilde{r}_{N-1}]^T = \mathbf{F}\mathbf{r} = \tilde{\mathbf{H}}\tilde{\mathbf{d}}_1 + \tilde{\mathbf{G}}\tilde{\mathbf{d}}_2 + \tilde{\mathbf{n}}_r, \qquad (4.3)$$

where $\tilde{\mathbf{H}}$ and $\tilde{\mathbf{G}}$ are the diagonal matrices with the diagonal elements given by the N-point DFT of $\mathbf{h}$ and $\mathbf{g}$, i.e.,

$$\tilde{\mathbf{h}} = \sqrt{N}\mathbf{F}_{L_1}\mathbf{h} = [\tilde{h}_0, \tilde{h}_1, \ldots, \tilde{h}_{N-1}]^T,$$

$$\tilde{\mathbf{g}} = \sqrt{N}\mathbf{F}_{L_2}\mathbf{g} = [\tilde{g}_0, \tilde{g}_1, \ldots, \tilde{g}_{N-1}]^T,$$

and $\mathbf{F}_x$ is defined as the matrix that contains the first x columns of $\mathbf{F}$. Moreover, $\tilde{\mathbf{n}}_r = [\tilde{n}_{r,0}, \tilde{n}_{r,1}, \ldots, \tilde{n}_{r,N-1}]^T$ is the corresponding AWGN vector in the frequency domain, each element with variance σ_n^2.

4.1.2 Relay Processing

The channel estimation of $\mathbf{h}, \mathbf{g}$ can be performed at $\mathbb{R}$ using the traditional method in MISO OFDM system [5,6], and will be omitted here.

During Phase II, the signal $\mathbf{u}$ being broadcasted by $\mathbb{R}$ is a linearly precoded version of $\mathbf{r}$, i.e.,

$$\mathbf{u} = [u_0, u_1, \ldots, u_{N-1}]^T = \mathbf{Ar}, \tag{4.4}$$

where $\mathbf{A}$ is the precoding matrix. To keep the original intention of adopting OFDM, $\mathbf{A}$ should be designed in a way that there is no inter-carrier interference (ICI). Therefore, $\mathbf{A}$ can be represented by

$$\mathbf{A} = \mathbf{F}^H \tilde{\mathbf{A}} \mathbf{J} \mathbf{F}, \tag{4.5}$$

where $\mathbf{J}$ is an permutation matrix that can be optimized according to [7], and $\tilde{\mathbf{A}}$ is a diagonal matrix defining the power allocation on each subcarrier.

Matrix $\mathbf{A}$ should keep the average time-domain symbol power constraint of $\mathbb{R}$ as $\mathrm{E}\{|u_i|^2\} \leq P_r$. Denote the diagonal elements of $\tilde{\mathbf{A}}$ as $\tilde{\mathbf{a}} = [\tilde{a}_0, \tilde{a}_1, \ldots, \tilde{a}_{N-1}]^T$. It can be readily verified that, this power constraint is equivalent to

$$\sum_{i=0}^{N-1} |\tilde{a}_i|^2 \leq \frac{NP_r}{\beta_1 P_1 + \beta_2 P_2 + \sigma_n^2}, \tag{4.6}$$

where $\beta_1 \triangleq \sum_{l=0}^{L_1-1} \sigma_{h,l}^2$ and $\beta_2 \triangleq \sum_{l=0}^{L_2-1} \sigma_{g,l}^2$ are introduced for notational simplicity.

Finally, $\mathbb{R}$ adds a new CP to $\mathbf{u}$ and broadcasts the resultant signal to both source terminals.

4.1.3   *Maximum Likelihood Data Detection*

Due to symmetry, we only illustrate the process at $\mathbb{T}_1$. In Phase II, $\mathbb{T}_1$ receives, after CP removal, the signal

$$\mathbf{y} = \mathbf{HAHd}_1 + \mathbf{HAGd}_2 + \mathbf{HAn}_r + \mathbf{n}_1, \tag{4.7}$$

where $\mathbf{n}_1 = [n_{1,0}, n_{1,1}, \dots, n_{1,N-1}]^T$ is the $N \times 1$ noise vector at $\mathbb{T}_1$ and has the same statistics as $\mathbf{n}_r$. The normalized DFT of $\mathbf{y}$ is

$$\tilde{\mathbf{y}} = \tilde{\mathbf{H}}\tilde{\mathbf{A}}\mathbf{J}\tilde{\mathbf{H}}\tilde{\mathbf{d}}_1 + \tilde{\mathbf{H}}\tilde{\mathbf{A}}\mathbf{J}\tilde{\mathbf{G}}\tilde{\mathbf{d}}_2 + \tilde{\mathbf{H}}\tilde{\mathbf{A}}\mathbf{J}\tilde{\mathbf{n}}_r + \tilde{\mathbf{n}}_1, \tag{4.8}$$

where $\tilde{\mathbf{n}}_1 = [\tilde{n}_{1,0}, \tilde{n}_{1,1}, \dots, \tilde{n}_{1,N-1}]^T$ is the corresponding AWGN in the frequency domain. Suppose the effect of left multiplying $\mathbf{J}$ on a vector is to permute its $\pi(i)$th element to the ith position. Then, $\tilde{\mathbf{y}}$ can be rewritten as

$$\tilde{\mathbf{y}} = \tilde{\mathbf{a}} \odot \tilde{\mathbf{b}} \odot \tilde{\mathbf{d}}_1 + \tilde{\mathbf{a}} \odot \tilde{\mathbf{c}} \odot \tilde{\mathbf{d}}_2 + \tilde{\mathbf{H}}\tilde{\mathbf{A}}\mathbf{J}\tilde{\mathbf{n}}_r + \tilde{\mathbf{n}}_1, \tag{4.9}$$

where

$$\tilde{\mathbf{b}} = \tilde{\mathbf{h}} \odot \tilde{\mathbf{h}}_\pi, \quad \tilde{\mathbf{c}} = \tilde{\mathbf{h}} \odot \tilde{\mathbf{g}}_\pi,$$

$$\tilde{\mathbf{h}}_\pi = [\tilde{h}_{\pi(0)}, \tilde{h}_{\pi(1)}, \dots, \tilde{h}_{\pi(N-1)}]^T,$$

$$\tilde{\mathbf{g}}_\pi = [\tilde{g}_{\pi(0)}, \tilde{g}_{\pi(1)}, \dots, \tilde{g}_{\pi(N-1)}]^T,$$

and $\odot$ denotes the Hadamard product between two vectors. The received signal on the ith carrier can be separately written as

$$\tilde{y}_i = \tilde{a}_i \tilde{h}_i \tilde{h}_{\pi(i)} \tilde{d}_{1,\pi(i)} + \tilde{a}_i \tilde{h}_i \tilde{g}_{\pi(i)} \tilde{d}_{2,\pi(i)} + \tilde{a}_i \tilde{h}_i \tilde{n}_{r,\pi(i)} + \tilde{n}_{1,i}. \tag{4.10}$$

Note that the permutation function $\pi(\cdot)$, or equivalently the matrix $\mathbf{J}$ must be known as a prior to $\mathbb{T}_1$ in order to decode the data.

Since $\mathbb{T}_1$ knows its own signal, the coherent ML data detection for $\tilde{d}_{2,\pi(i)}$ is obtained from the conditional probability density function (PDF) $p(\tilde{y}_i | \tilde{d}_{2,i}, \tilde{\mathbf{h}}, \tilde{\mathbf{g}})$ as

$$
\begin{aligned}
\hat{\tilde{d}}_{2,\pi(i)} &= \arg\max_{x \in \mathcal{C}_2} \frac{1}{\pi \sigma_n^2 (|\tilde{h}_i|^2 |\tilde{a}_i|^2 + 1)} \exp\left\{ -\frac{|\tilde{y}_i - \tilde{a}_i \tilde{h}_i \tilde{h}_{\pi(i)} \tilde{d}_{1,\pi(i)} - \tilde{a}_i \tilde{h}_i \tilde{g}_{\pi(i)} x|^2}{(|\tilde{h}_i|^2 |\tilde{a}_i|^2 + 1) \sigma_n^2} \right\} \\
&= \arg\min_{x \in \mathcal{C}_2} |\tilde{y}_i - \tilde{a}_i \tilde{h}_i \tilde{h}_{\pi(i)} \tilde{d}_{1,\pi(i)} - \tilde{a}_i \tilde{h}_i \tilde{g}_{\pi(i)} x|^2,
\end{aligned} \tag{4.11}
$$

where $\mathcal{C}_2$ is the signal constellation of $\mathbb{T}_2$.

Remark. Note that, only the cascaded channels $\tilde{h}_i \tilde{h}_{\pi(i)}$ and $\tilde{h}_i \tilde{g}_{\pi(i)}$ are needed for data detection.

4.1.4 Channel Estimation Strategy

Although the knowledge of the individual channels $\mathbf{h}$ and $\mathbf{g}$ do not contribute to the ML data detection, the task of the channel estimation in PLNC should still focus on estimating $\mathbf{h}$ and $\mathbf{g}$, as mentioned in Chap. 2:

1. Knowing $\mathbf{h}$ and $\mathbf{g}$ at $\mathbb{T}_1$ can help $\mathbb{T}_1$ predict the relay operation $\pi(\cdot)$ and thus eliminates the necessity of the relay's feedback.
2. If the permutation $\pi(\cdot)$ is changed by $\mathbb{R}$, $\mathbb{T}_1$ does not need to re-estimate the cascaded channels. Instead, it can construct the new cascaded channels once knowing the new permutation function.

Definition 4.1. For an $N \times 1$ vector $\mathbf{x} = [x_0, x_1, \ldots, x_{N-1}]^T$, if $x_{L-1} \neq 0$ but $x_i = 0$ for $\forall i \geq L$, we say $\mathbf{x}$ has size N but has length L.

Further more, define the N-point IDFT (not normalized) of $\tilde{\mathbf{h}}_\pi$, $\tilde{\mathbf{g}}_\pi$, $\tilde{\mathbf{b}}$, and $\tilde{\mathbf{c}}$ as $\mathbf{h}_\pi$, $\mathbf{g}_\pi$, $\mathbf{b}$, and $\mathbf{c}$ respectively. Then,

$$\mathbf{b} = \mathbf{h} \otimes \mathbf{h}_\pi, \quad \text{and} \quad \mathbf{c} = \mathbf{h} \otimes \mathbf{g}_\pi$$

are the cascaded channels in the time domain, where $\otimes$ denotes the N-point circular convolution.

The permutation matrix $\mathbf{J}$ during the data transmission should be obtained according to certain optimal criterion. Nevertheless, we could use $\mathbf{J} = \mathbf{I}$ during training due to the following reasons:

1. If $\mathbf{J} = \mathbf{I}$, then the lengths of $\mathbf{h}_\pi = \mathbf{h}$ and $\mathbf{g}_\pi = \mathbf{g}$ reach their minimum L_1 and L_2, respectively. Otherwise, $\mathbf{h}_\pi$ and $\mathbf{g}_\pi$ will in general have lengths N due to the randomness of channels.[2] Consequently, the lengths of $\mathbf{b}$ and $\mathbf{c}$ reach their minimum, namely, $2L_1 - 1$ and $L_1 + L_2 - 1$.
2. Selecting $\mathbf{J} = \mathbf{I}$ could help to separately retrieve $\mathbf{h}$ and $\mathbf{g}$ from $\mathbf{b}$ and $\mathbf{c}$, as will be seen later.

With $\mathbf{J} = \mathbf{I}$, (4.7) becomes

$$\mathbf{y} = \mathbf{F}^H \tilde{\mathbf{A}} \tilde{\mathbf{D}}_1 \tilde{\mathbf{b}} + \mathbf{F}^H \tilde{\mathbf{A}} \tilde{\mathbf{D}}_2 \tilde{\mathbf{c}} + \mathbf{H} \mathbf{A} \mathbf{n}_r + \mathbf{n}_1$$
$$= \mathbf{A} \mathbf{D}_1 \mathbf{b} + \mathbf{A} \mathbf{D}_2 \mathbf{c} + \mathbf{H} \mathbf{A} \mathbf{n}_r + \mathbf{n}_1, \qquad (4.12)$$

where $\tilde{\mathbf{D}}_j = \text{diag}\{\tilde{\mathbf{d}}_j\}$, $j = 1, 2$, is the $N \times N$ diagonal matrix, $\mathbf{D}_1$ is the $N \times (2L_1 - 1)$ column-wise circulant matrix with the first column $\mathbf{d}_1$, and $\mathbf{D}_2$ is the $N \times (L_1 + L_2 - 1)$ column-wise circulant matrix with the first column $\mathbf{d}_2$.

---

[2]We assume that $N \geq \max\{2L_1 - 1, L_1 + L_2 - 1\}$.

4.1.5   Specialities of PLNC

If the frequency domain cascaded channels $\tilde{\mathbf{b}}$ and $\tilde{\mathbf{c}}$ are estimated in the first place, the number of unknowns is $2N$. However, if the time domain cascaded channels $\mathbf{b}$ and $\mathbf{c}$ are estimated in the first place, the number of unknowns reduces to $3L_1 + L_2 - 2$. Due to this large saving, practical OFDM systems usually focus on estimating the time-domain channel vector [4–6]. In fact, first estimating $\mathbf{b}$ and $\mathbf{c}$ still results in much redundancy in terms of the number of degrees of freedom, because $\mathbf{h}$ and $\mathbf{g}$ together have $L_1 + L_2$ unknown parameters. Hence, a critical issue for channel estimation in PLNC system is to reduce the account of the training to its least value.

Meanwhile, whether $\mathbf{h}$ and $\mathbf{g}$ can be uniquely recovered from $\mathbf{b}$ and $\mathbf{c}$ is also a critical issue for PLNC, i.e., the identifiability problem.

Both the above issues represent the specialities of channel estimation in PLNC.

4.2   Block Training Based Channel Estimation

Using a whole OFDM block for training normally happens at the start of the transmission, and this OFDM block is known as the preamble. In this section, we first estimate $\mathbf{b}$ and $\mathbf{c}$ and then recover $\mathbf{h}$ and $\mathbf{g}$ from a specifically designed algorithm.

4.2.1   Channel Estimation

The least-square (LS) estimates of $\mathbf{b}$ and $\mathbf{c}$ are obtained from

$$[\hat{\mathbf{b}}^T, \hat{\mathbf{c}}^T]^T = \arg\min_{\mathbf{b},\mathbf{c}} \|\mathbf{y}_1 - \mathbf{A}\mathbf{D}_1\mathbf{b} - \mathbf{A}\mathbf{D}_2\mathbf{c}\|^2 = (\mathbf{D}^H\mathbf{A}^H\mathbf{A}\mathbf{D})^{-1}\mathbf{D}^H\mathbf{A}^H\mathbf{y}_1,$$

$$(4.13)$$

where $\mathbf{D} \triangleq [\mathbf{D}_1 \ \mathbf{D}_2]$. The corresponding error covariance is

$$\mathrm{Cov}\{\Delta\mathbf{e}\} = \sigma_n^2 (\mathbf{D}^H\mathbf{A}^H\mathbf{A}\mathbf{D})^{-1} \left(\mathbf{D}^H\mathbf{A}^H \left(\beta_1\mathbf{A}\mathbf{A}^H + \mathbf{I}\right)\mathbf{A}\mathbf{D}\right) (\mathbf{D}^H\mathbf{A}^H\mathbf{A}\mathbf{D})^{-1}.$$

$$(4.14)$$

4.2.2   Training Sequence Design

The optimal training sequence $\tilde{\mathbf{d}}_j$'s and the optimal relay precoding matrix $\mathbf{A}$ should be found by minimizing channel estimation MSE. Unfortunately, the closed-form solutions are currently unavailable due to the complicated structure of $\mathrm{Cov}\{\Delta\mathbf{e}\}$.

Nevertheless, since we have the freedom in designing $\mathbf{A}$, we can simplify the problem by considering $\tilde{\mathbf{A}}$ as a scaled identity matrix, which is intuitively reasonable since carriers are symmetric among themselves in terms of channel gain and noise power. Then the MSE becomes

$$\text{MSE} = tr(\text{Cov}\{\Delta\mathbf{e}\}) = \sigma_n^2\left(\beta_1 + \frac{\beta_1 P_1 + \beta_2 P_2 + \sigma_n^2}{P_r}\right) tr\left((\mathbf{D}^H\mathbf{D})^{-1}\right),$$
(4.15)

and the optimal $\tilde{\mathbf{d}}_j$'s should minimize $tr\left((\mathbf{D}^H\mathbf{D})^{-1}\right)$.

Although the problem looks similar to that in a 2×1 MISO OFDM system, the traditional way [5, 6] cannot be directly applied since the two terminals have their individual power constraints: $\|\mathbf{d}_1\|^2 = NP_1$ and $\|\mathbf{d}_2\|^2 = NP_2$.[3]

Theorem 4.1. *The optimal training $\tilde{\mathbf{d}}_j$'s that minimize (4.15) should satisfy*

(C1) $\quad |\tilde{d}_{j,i}|^2 = P_j, \quad \forall j = 1, 2, \ i = 0, \ldots, N - 1,$

(C2) $\quad \displaystyle\sum_{i=0}^{N-1} \tilde{d}_{1,i}^* \tilde{d}_{2,i} e^{-j2\pi ki/N} = 0,$

$$\forall k \in \{-(L_1 + L_2 - 2), -(L_1 + L_2 - 3), \ldots, (2L_1 - 2)\}.$$

Proof. See Appendix "Proof of Theorem 4.1". □

With the optimal training, the estimation MSEs of $\mathbf{b}$ and $\mathbf{c}$ also reach their individual minimum as

$$\text{MSE}(\mathbf{b}) = \sigma_n^2\left(\beta_1 + \frac{\beta_1 P_1 + \beta_2 P_2 + \sigma_n^2}{P_r}\right)\frac{2L_1 - 1}{NP_1}, \tag{4.16a}$$

$$\text{MSE}(\mathbf{c}) = \sigma_n^2\left(\beta_1 + \frac{\beta_1 P_1 + \beta_2 P_2 + \sigma_n^2}{P_r}\right)\frac{L_1 + L_2 - 1}{NP_2}. \tag{4.16b}$$

Remark. Due to symmetry in the structure of $\tilde{\mathbf{d}}_j$'s, the optimal channel estimation could be achieved for two terminals simultaneously with the same set of $\tilde{\mathbf{d}}_j$'s.

4.2.3 Identifiability and Individual Channel Extraction

To make sure that $\mathbf{h}$ and $\mathbf{g}$ can be uniquely recovered from $\mathbf{b}$ and $\mathbf{c}$, we study the channel identifiability in this subsection.

---

[3]In traditional works, there is only one total power constraint $\|\mathbf{d}_1\|^2 + \|\mathbf{d}_1\|^2 \leq P$, for some power value P.

Theorem 4.2. *If L_1 is known and $L_1 < N$, then $\mathbf{h}$ can be determined from $\mathbf{b}$ with sign ambiguity.*

Proof. See Appendix "Proof of Theorem 4.2".   □

Corollary 4.1. *Once $\mathbf{h}$ or $-\mathbf{h}$ is found, $\mathbf{g}$ or $-\mathbf{g}$ can be determined from $\mathbf{c}$.*

Theorem 4.3. *If L_2 is known and $L_2 < N$, then $\mathbf{g}$ can be determined from $\mathbf{b}$ and $\mathbf{c}$ with sign ambiguity.*

Proof. See Appendix "Proof of Theorem 4.3".   □

Remark.. The sign ambiguity in $\mathbf{h}$ and $\mathbf{g}$ appears only in pair as $\{-\mathbf{h}, -\mathbf{g}\}$. This scenario is called simultaneous sign ambiguity (SSA). It is known from (4.9) that SSA does not affect the detection at source terminals, nor does it affect the prediction of relay actions, e.g., carrier permutation and power allocation [7].

Definition 4.2. If only SSA exists for the estimated $\mathbf{h}$ and $\mathbf{g}$, we claim that the channel is identifiable.

In practical scenario, only the upper bound of L_j, $j = 1, 2$, denoted as $\hat{L}_j$ can be obtained.

Theorem 4.4. *The maximum $\hat{L}_1$ to guarantee the channel identifiability is $\min\{L_1 + \frac{N}{2} - 1, N - 1\}$ for even N.*

Proof. See Appendix "Proof of Theorem 4.4".   □

Corollary 4.2. *The maximum $\hat{L}_2$ to guarantee the channel identifiability is $\min\{L_2 + \frac{N}{2} - 1, N - 1\}$ for even N.*

Theorem 4.5. *For odd N, $\hat{L}_1 \leq \min\{L_1 + \frac{N-1}{2}, N - 1\}$ can guarantee the channel identifiability.*

Proof. See Appendix "Proof of Theorem 4.5".   □

For both even and odd N, the practical channel order overestimate $(\hat{L}_1 - L_1)$ is much less than $N/2$, e.g., in IEEE 802.11a [8], $\hat{L}_1 \leq L_{cp} \leq N/4$. Therefore, the order overestimation, in most cases, does not cause the identifiability problem.

Remark. Although the identifiability can be guaranteed from Theorems 4.4 and 4.5, ambiguous results may still be picked up in the noisy environment.[4] This phenomenon is conventionally named as the *outlier*, which normally happens at low signal-to-noise ratio (SNR) region, as will be seen in the later simulation.

Lastly, we need to design algorithm to extract $\mathbf{h}$ and $\mathbf{g}$ from $\mathbf{b}$ and $\mathbf{c}$. Please see Appendix "Algorithm to Extract $\mathbf{h}$ and $\mathbf{g}$" for details.

---

[4]We need to mention that the identifiability is a technical term defined for noiseless environment.

4.3 Pilot Tone Based Channel Estimation

In many OFDM standards, several PTs reside in one OFDM block for tracking parameters like channel coefficients, frequency offset, phase noises, etc [5, 6, 9]. To save the bandwidth, the amount of PTs should be kept as small as possible. In this section, we are interested in finding the minimum required number of PTs, as well as designing the corresponding channel estimation algorithms.

To simplify the design, we propose $\mathbb{T}_1$ and $\mathbb{T}_2$ to send pilots on non-overlapping carriers with indices $\mathcal{L}_1 = \{i_0, i_1, \ldots, i_{(K_1-1)}\}$ and $\mathcal{L}_2 = \{q_0, q_1, \ldots, q_{(K_2-1)}\}$, respectively, i.e., $\mathcal{L}_1 \bigcap \mathcal{L}_2 = \phi$. To avoid the interference, $\mathbb{T}_1$ should assign zero values to carriers with indices in $\mathcal{L}_2$, while $\mathbb{T}_2$ should assign zero values to carriers in $\mathcal{L}_1$. Therefore, the total number of carriers reserved for training from either terminal is $K_1 + K_2$.

4.3.1 Channel Estimation

1. *LS estimation of* $\tilde{b}_{i_k}$: The received signal on carrier $i_k \in \mathcal{L}_1$ is written as

$$\tilde{y}_{i_k} = \tilde{a}_{i_k} \tilde{d}_{1,i_k} \tilde{h}_{i_k}^2 + \tilde{h}_{i_k} \tilde{a}_{i_k} \tilde{n}_{r,i_k} + \tilde{n}_{1,i_k}, \quad i_k \in \mathcal{L}_1, \qquad (4.17)$$

from which the LS estimation of $\tilde{b}_{i_k}$ is expressed as

$$\hat{\tilde{b}}_{i_k} = \frac{\tilde{y}_{i_k}}{\tilde{a}_{i_k} \tilde{d}_{1,i_k}} = \tilde{b}_{i_k} + \frac{\tilde{h}_{i_k} \tilde{n}_{r,i_k}}{\tilde{d}_{1,i_k}} + \frac{\tilde{n}_{1,i_k}}{\tilde{a}_{i_k} \tilde{d}_{1,i_k}}. \qquad (4.18)$$

2. *LS estimation of* $\tilde{c}_{q_k}$: The received signals on carrier $q_k \in \mathcal{L}_2$ is written as

$$\tilde{y}_{q_k} = \tilde{a}_{q_k} \tilde{d}_{2,q_k} \tilde{h}_{q_k} \tilde{g}_{q_k} + \tilde{h}_{q_k} \tilde{a}_{q_k} \tilde{n}_{r,q_k} + \tilde{n}_{1,q_k}, \quad q_k \in \mathcal{L}_2, \qquad (4.19)$$

from which the LS estimation of $\tilde{c}_{i_k}$ is expressed as

$$\hat{\tilde{c}}_{q_k} = \frac{\tilde{y}_{q_k}}{\tilde{a}_{q_k} \tilde{d}_{2,q_k}} = \tilde{c}_{q_k} + \frac{\tilde{h}_{q_k} \tilde{n}_{r,q_k}}{\tilde{d}_{2,q_k}} + \frac{\tilde{n}_{1,q_k}}{\tilde{a}_{q_k} \tilde{d}_{2,q_k}}. \qquad (4.20)$$

4.3.2 Identifiability and Individual Channel Extraction

Consider the noise-free case, where $\mathbb{T}_1$ can perfectly estimate $\tilde{\mathbf{b}}' = [\tilde{b}_{i_0}, \tilde{b}_{i_1}, \ldots, \tilde{b}_{i_{(K_1-1)}}]^T$ through carrier set $\mathcal{L}_1$ while estimate $\tilde{\mathbf{c}}' = [\tilde{c}_{q_0}, \tilde{c}_{q_1}, \ldots, \tilde{c}_{q_{(K_2-1)}}]^T$ through carrier set $\mathcal{L}_2$.

Theorem 4.6. *If $K_1 > L_1$ and the indices in $\mathcal{L}_1$ are equispaced,[5] then $\mathbf{h}$ can be determined from $\tilde{\mathbf{b}}'$ with sign ambiguity only.*

Proof. See Appendix "Proof of Theorem 4.6".   □

Theorem 4.6 can be considered as a counterpart of Theorem 4.2 but is defined on a limited number of PTs.

Conjecture 4.1. *If $K_1 > L_1$, $\mathbf{h}$ can be determined from $\tilde{\mathbf{b}}'$ with sign ambiguity only.*

See heuristic discussions in Appendix "Illustration of Conjecture 4.1". Conjecture 4.1 has been verified by numerous simulations in noiseless environment.

Theorem 4.7. *When $\mathcal{L}_1$ satisfies conditions in Theorem 4.6, if $K_2 \geq L_2$, then $\mathbf{g}$ can be determined with sign ambiguity only.*

Proof. See Appendix "Proof of Theorem 4.7".   □

In practical scenario when the channel length is overestimated, we have the following results:

Theorem 4.8. *If $\mathcal{L}_1$ contains K_1 equispaced indices, the maximum $\hat{L}_1$ to guarantee only the sign ambiguity is $\min\{L_1 + \frac{K_1}{2} - 1, K_1 - 1\}$ for even K_1.*

Proof. The proof directly follows that of Theorem 4.4.   □

Remark. From the above discussions, the minimum number of PTs in one OFDM block should be $(L_1 + 1) + L_2$, where $L_1 + 1$ PTs are used by $\mathbb{T}_1$ and L_2 PTs are used by $\mathbb{T}_2$. Now considering also the channel estimation at $\mathbb{T}_2$. It then requires that $L_2 + 1$ PTs are used by $\mathbb{T}_2$ and L_1 PTs are used by $\mathbb{T}_1$. Therefore, for the whole PLNC, where channel estimations must be performed at both terminals, $\mathbb{T}_1$ should occupy at least $L_1 + 1$ PTs and $\mathbb{T}_2$ should occupy at least $L_2 + 1$ PTs. The minimum number of PTs in one OFDM block should be $(L_1 + 1) + (L_2 + 1)$.

Lastly, we need to design algorithm to extract $\mathbf{h}$ and $\mathbf{g}$ from $\tilde{\mathbf{b}}'$ and $\tilde{\mathbf{c}}'$. Please see Appendix "Efficient Algorithm to Estimate $\mathbf{h}$ and $\mathbf{g}$" for details.

4.3.3 Pilot Tone Design

1. Training Design at $\mathbb{T}_1$ and $\mathbb{R}$:

In addition to PT selection and power allocation at $\mathbb{T}_1$, we should also design the scaling factor $\tilde{a}_{i,k}$ at $\mathbb{R}$. Since PTs and data carriers co-exist in one OFDM block, then the constraint (4.6) should be modified. For the time being, we will

---

[5]Being equispaced implies that K_1 divides N.

apply a modified constraint as $\sum_{k=0}^{K_1-1} |\tilde{a}_{i_k}|^2 \leq P_{r_{t1}}$, $i_k \in \mathcal{L}_1$. Meanwhile, the training power constraint from $\mathbb{T}_1$ is $\sum_{k=0}^{K_1-1} |\tilde{d}_{i_k}|^2 \leq P_{1_t}$, where P_{1_t} is the total power assigned for PTs at $\mathbb{T}_1$.

For the LS estimation in (4.18), the error cross-correlation of the estimation, conditioned on a specific realization of $\mathbf{h}$, is

$$\mathrm{E}\{(\hat{\tilde{b}}_{i_k} - \tilde{b}_{i_k})(\hat{\tilde{b}}_{i_j} - \tilde{b}_{i_j})^*|\mathbf{h}\} = \left(\frac{\sigma_n^2 |\tilde{h}_{i_k}|^2}{|\tilde{d}_{1,i_k}|^2} + \frac{\sigma_n^2}{|\tilde{a}_{i_k} \tilde{d}_{1,i_k}|^2} \right) \delta(k-j), \qquad (4.21)$$

where $\delta(\cdot)$ is the delta function that is one when $k = j$ but is zero otherwise. Since $\hat{\tilde{h}}_{i_k}$ is obtained from $\tilde{b}_{i_k}^{1/2}$, (4.21) can be rewritten as

$$\mathrm{E}\{((\tilde{h}_{i_k} + \Delta\tilde{h}_{i_k})^2 - \tilde{h}_{i_k}^2)((\tilde{h}_{i_j} + \Delta\tilde{h}_{i_j})^2 - \tilde{h}_{i_j}^2)^*|\mathbf{h}\} \approx \mathrm{E}\{4\tilde{h}_{i_k} \tilde{h}_{i_j}^* \Delta\tilde{h}_{i_k} \Delta\tilde{h}_{i_j}^*|\mathbf{h}\}, \qquad (4.22)$$

where the RHS is obtained by discarding the higher order statistics of $\Delta\tilde{h}_{i_k}$ and $\Delta\tilde{h}_{i_j}$. Then

$$\mathrm{E}\{\Delta\tilde{h}_{i_k} \Delta\tilde{h}_{i_j}^*|\mathbf{h}\} \approx \left(\frac{\sigma_n^2}{4|\tilde{d}_{1,i_k}|^2} + \frac{\sigma_n^2}{4|\tilde{h}_{i_k}|^2 |\tilde{a}_{i_k} \tilde{d}_{1,i_k}|^2} \right) \delta(k-j). \qquad (4.23)$$

Define $\Delta\tilde{\mathbf{h}} = [\Delta\tilde{h}_{i_0}, \Delta\tilde{h}_{i_1}, \ldots, \Delta\tilde{h}_{i_{(K_1-1)}}]^T$ and $\Delta\mathbf{h} = [\Delta h_0, \Delta h_1, \ldots, \Delta h_{L-1}]^T$. The covariance matrix of $\Delta\tilde{\mathbf{h}}$ is

$$\mathrm{Cov}(\Delta\tilde{\mathbf{h}}) = \mathrm{diag}\left\{ \mathrm{E}\{|\Delta\tilde{h}_{i_0}|^2\}, \ldots, \mathrm{E}\{|\Delta\tilde{h}_{i_{(K_1-1)}}|^2\} \right\}, \qquad (4.24)$$

and the covariance matrix of $\Delta\mathbf{h}$ is

$$\mathrm{Cov}(\Delta\mathbf{h}) = \frac{1}{N} \mathbf{F}_{K_1 L_1}^\dagger \mathrm{Cov}(\Delta\tilde{\mathbf{h}})(\mathbf{F}_{K_1 L_1}^\dagger)^H, \qquad (4.25)$$

where $\mathbf{F}_{K_1 L_1}$ is defined in Appendix "Proof of Theorem 4.6".

The optimal PT can then be found from

$$\min_{\tilde{a}_{i_k}, \tilde{d}_{i_k}, \mathcal{L}_1} \quad \mathrm{MSE} = tr((\mathbf{F}_{K_1 L_1} \mathbf{F}_{K_1 L_1}^H)^\dagger \mathrm{Cov}(\Delta\tilde{\mathbf{h}})) \qquad (4.26)$$

$$\text{s.t.} \quad \sum_{k=1}^{K_1} |\tilde{a}_{i_k}|^2 \leq P_{r_{t1}}, \quad \sum_{k=1}^{K_1} |\tilde{d}_{i_k}|^2 \leq P_{1_t}.$$

The above optimization is hard to solve due to its complicated structure. We here propose to minimize the upper bound of the MSE. Define $M_\alpha = \max\limits_{i_k \in \mathcal{L}_1} \dfrac{\sigma_n^2}{4|\tilde{d}_{1,i_k}|^2} +$
$\dfrac{\sigma_n^2}{4|\tilde{h}_{i_k}|^2|\tilde{a}_{i_k}\tilde{d}_{1,i_k}|^2}$. Since $\mathrm{Cov}(\Delta\tilde{\mathbf{h}})$ is a diagonal matrix, there is

$$\mathrm{MSE} \leq M_\alpha tr((\mathbf{F}_{K_1 L_1}\mathbf{F}_{K_1 L_1}^H)^\dagger) = M_\alpha tr((\mathbf{F}_{K_1 L_1}^H \mathbf{F}_{K_1 L_1})^{-1}). \tag{4.27}$$

Now the subcarrier index selection becomes the same as that in the traditional work [4], and the optimal K_1 should divide N and the optimal i_k's should be equispaced.

Substituting the corresponding $\mathbf{F}_{K_1 L_1}$ into (4.26) gives the new objective function:

$$\min_{\tilde{a}_{i_k},\tilde{d}_{i_k}} \quad \frac{L_1\sigma_n^2}{4K_1^2}\sum_{k=1}^{K_1}\frac{1}{|\tilde{d}_{1,i_k}|^2} + \frac{1}{|\tilde{h}_{i_k}|^2|\tilde{a}_{i_k}\tilde{d}_{1,i_k}|^2}. \tag{4.28}$$

Finally, we need to deal with the unknown $|\tilde{h}_{i_k}|^2$ in the denominator. Considering $\tilde{h}_{i_k} = \sum_{l=0}^{L_1-1} h_l e^{-j2\pi i_k l/N}$, $|\tilde{h}_{i_k}|^2$ can be represented by its expectation $\mathrm{E}\{|\tilde{h}_{i_k}|^2\} = \beta_1$ if L_1 is relatively large.

The final optimization problem to be solved is

$$\min_{\tilde{a}_{i_k},\tilde{d}_{1,i_k}} \quad \sum_{k=0}^{K_1-1}\frac{1}{|\tilde{d}_{1,i_k}|^2} + \frac{1}{\beta_1|\tilde{a}_{i_k}\tilde{d}_{1,i_k}|^2} \tag{4.29}$$

$$\text{s.t.} \quad \sum_{k=0}^{K_1-1}|\tilde{a}_{i_k}|^2 \leq P_{r_{t1}}, \quad \sum_{k=0}^{K_1-1}|\tilde{d}_{1,i_k}|^2 \leq P_{1_t}.$$

The problem is standard convex optimization in terms of $|\tilde{d}_{1,i_k}|^2$ and $|\tilde{\alpha}_{i_k}|^2$, and can be solved from the Karush–Kuhn–Tucker (KKT) conditions [10]. The optimal solution can be obtained as $|\tilde{d}_{1,i_k}|^2 = P_{1_t}/K_1$ and $|\tilde{\alpha}_{i_k}|^2 = P_{r_{t1}}/K_1$.

2. *Training Design at* $\mathbb{T}_2$ *and* $\mathbb{R}$:

Similarly to previous subsection, we apply a modified scaling constraint over $\mathcal{L}_2$ as $\sum_{k=0}^{K_2-1}|\tilde{a}_{qk}|^2 \leq P_{r_{t2}}$, From (4.20), the error cross-correlation of the estimation, conditioned on a specific realization of $\mathbf{h}$, is

$$\mathrm{E}\{(\hat{\tilde{c}}_{qk} - \tilde{c}_{qk})(\hat{\tilde{c}}_{qj} - \tilde{c}_{qj})^*|\mathbf{h}\}$$

$$= \mathrm{E}\{(\tilde{h}_{qk}\Delta\tilde{g}_{qk})(\tilde{h}_{qj}\Delta\tilde{g}_{qj})^*|\mathbf{h}\} = \left(\frac{\sigma_n^2|\tilde{h}_{qk}|^2}{|\tilde{d}_{2,qk}|^2} + \frac{\sigma_n^2}{|\tilde{a}_{qk}\tilde{d}_{2,qk}|^2}\right)\delta(k-j),$$

$$\tag{4.30}$$

where we do not consider the channel estimation error in $\tilde{h}_{q_k}$, $q_k \in \mathcal{L}_2$ for the time being due to its non-extractable MSE expression. Following the similar approach from (4.21) to (4.29), we arrive at

$$
\min_{\tilde{a}_{q_k}, \tilde{d}_{2,q_k}} \sum_{k=0}^{K_2-1} \frac{1}{|\tilde{d}_{2,q_k}|^2} + \frac{1}{\beta_1 |\tilde{a}_{q_k} \tilde{d}_{2,q_k}|^2} \tag{4.31}
$$

$$
\text{s.t.} \quad \sum_{k=0}^{K_2-1} |\tilde{a}_{q_k}|^2 \leq P_{r_{t2}}, \quad \sum_{k=0}^{K_2-1} |\tilde{d}_{2,q_k}|^2 \leq P_{2_t},
$$

where P_{2_t} is the training power that could be afforded at $\mathbb{T}_2$. The optimal solution can be obtained as $|\tilde{d}_{2,q_k}|^2 = P_{2_t} / K_2$ and $|\tilde{\alpha}_{q_k}|^2 = P_{r_{t2}} / K_2$.

Remark. Importantly, the above two separate approaches result in the consistent conclusions that equipowered and equispaced PTs should be applied, and hence a complicated joint design at all terminals is not needed.

Remark. Due to symmetry of the equipowered and equispaced PTs, the optimal channel estimation could be achieved for two terminals simultaneously with the same set of the pilot tones.

4.4 Numerical Results

In this section, we numerically examine the proposed studies over various scenarios. Both $\mathbf{h}$ and $\mathbf{g}$ have five taps, each with zero mean and unit variance. We always fix $P_1 = P_2 = P_r$ and define SNR as P_1/σ_n^2. The OFDM block length, following IEEE 802.11a, is set as $N = 64$, and the length of the CP is $L_{cp} = 16$. Totally 10^4 Monte-Carlo runs are adopted for averaging. The figure of the merit is the average MSE. For simulation purpose, the sign ambiguity in $\mathbf{h}$ is removed by choosing $\pm\hat{\mathbf{h}}$ from $\min\{\| + \hat{\mathbf{h}} - \mathbf{h}\|^2, \| - \hat{\mathbf{h}} - \mathbf{h}\|^2\}$ [11].

4.4.1 Training with One OFDM Block

1. Optimal $\tilde{\mathbf{d}}_j$'s versus non-optimal $\tilde{\mathbf{d}}_j$'s
    We then manually construct two non-optimal training sequences:

Type 1: $\tilde{d}_{j,i}$, $1 \leq i \leq 32$ is scaled by $\sqrt{0.5}$ from their optimal values while $\tilde{d}_{j,i}$, $33 \leq i \leq 64$ is scaled by a factor of $\sqrt{1.5}$ from their optimal values.
Type 2: $\tilde{d}_{j,i}$, $1 \leq i \leq 32$ is scaled by $\sqrt{0.1}$ from their optimal values while $\tilde{d}_{j,i}$, $33 \leq i \leq 64$ is scaled by a factor of $\sqrt{1.9}$ from their optimal values.

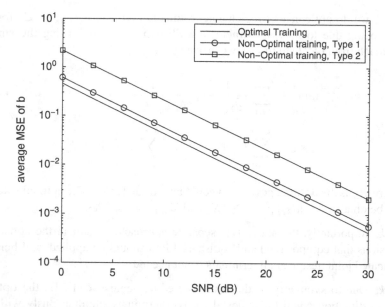

Fig. 4.2 Comparison for different training sequences in block-based channel estimation

We will then examine the proposed algorithm with the optimal training and the Type 1, Type 2 non-optimal training sequences. Clearly, all three types of training sequences have the same total power. Moreover, Type 1 training is closer to the optimal one than Type 2 training. The average MSEs of the estimated $\mathbf{b}$ versus SNR of different types of training are shown in Fig. 4.2. It is seen that as the training $\tilde{\mathbf{d}}_j$'s move farther from the optimal one, the channel estimation performance decreases. It needs to be mentioned that the average MSE of estimated $\mathbf{c}$ is the same as that of $\mathbf{b}$ due to the symmetric channel and training.

2. *Uniform $\tilde{a}_k$'s versus non-uniform $\tilde{a}_k$'s*

We then make the comparison by changing the values of $\tilde{\mathbf{a}}$. Two different types of non-uniform $\tilde{a}_k$'s are considered, whose formats follow the non-optimal choices in previous example. The average MSEs of the estimate of $\mathbf{b}$ versus SNR for these three types of scaling are shown in Fig. 4.3. Again, it is seen that as the scaling $\tilde{a}_k$'s move farther from the uniform one, the channel estimation performance degrades more severely.

3. *Recovering $\mathbf{h}$ and $\mathbf{g}$*

In this example, we recover $\mathbf{h}$ and $\mathbf{g}$ from $\mathbf{b}$ and $\mathbf{c}$, and display their average MSEs in Fig. 4.4. We then rebuild $\mathbf{b}$ and $\mathbf{c}$ from the estimated $\mathbf{h}$ and $\mathbf{c}$ whose average MSEs are also displayed in the same figure. To make a more fruitful comparison, we also include the average MSEs of $\mathbf{h}$ and $\mathbf{g}$ that are estimated at relays in the end of Phase I. Several observations are made here:

- The recovered $\mathbf{b}$ and $\mathbf{c}$ are more accurate than the directly estimated ones and this is called as *denoising effect*.

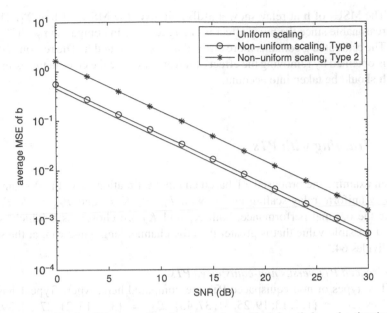

Fig. 4.3 Comparison for different scaling factors at relay in block-based channel estimation

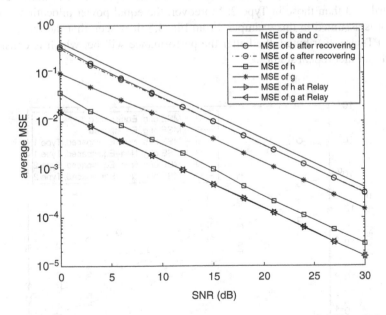

Fig. 4.4 Recovering the individual channels and reconstructing the cascaded channels in block-based channel estimation

- The MSEs of **h** at relay show a 3-dB gain over the MSEs of **h** at $\mathbb{T}_1$. This is reasonable since the noise effect is enlarged at $\mathbb{T}_1$ in comparison with $\mathbb{R}$.
- The estimated **g** contains more errors than the estimated **h**. The reason is that **g** is obtained by removing the effect of **h** from **c**, where the error in the estimated **h** should be taken into account.

4.4.2 Training with PTs

We then examine the proposed PT based channel estimation here. In all examples, we apply uniform relay scaling factors with $P_{r_{t1}} = K_1 P_r$ and $P_{r_{t2}} = K_2 P_r$. To achieve the optimal performance, both K_1 and K_2 are chosen as 8 which is the smallest possible value that is greater than the channel length and can, at the same time, divides 64.

1. Equispaced PTs versus non-equispaced PTs

Two types of non-equispaced PTs are compared here, where Type 1 has PT indices $\mathcal{L}_1 = \{1, 7, 13, 19, 25, 31, 37, 43\}$, $\mathcal{L}_2 = \{3, 9, 15, 21, 27, 33, 39, 45\}$ and Type 2 has PT indices $\mathcal{L}_1 = \{1, 6, 11, 16, 21, 26, 31, 36\}$, $\mathcal{L}_2 = \{3, 8, 13, 18, 23, 28, 33, 38\}$. Obviously, PTs in Type 1 are more evenly distributed than those in Type 2. Moreover, the equal power allocation to each pilot is assumed in this example. From Fig. 4.5, it is seen that the more evenly the PTs are distributed, the better the performance will be, which is consistent with our theoretical studies.

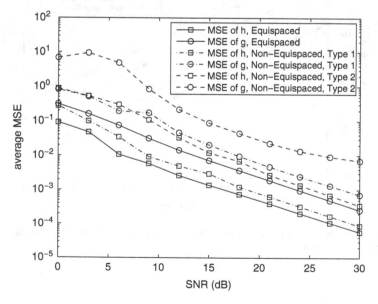

Fig. 4.5 Channel estimation under differently spaced PTs

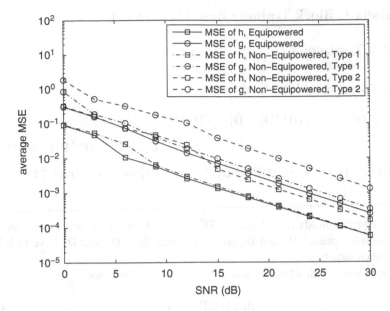

Fig. 4.6 Channel estimation under differently powered PTs

2. Equipowered PTs versus non-equipowered PTs

Two types of power allocation that follow the similar manner as Fig. 4.3 are selected; namely, power is more evenly allocated to different PTs in Type 1 than in Type 2. The equispaced PTs are adopted in this example. Again, it is observed from Fig. 4.6 that the more evenly the power is allocated, the better the channel estimation will be.

4.5 Summary

In this chapter, we proposed channel estimation algorithms under two different training schemes, i.e., block-based training and PT-based training. For both types of training, we analyzed the channel identifiability, proposed the LS-based channel estimators, and derived the optimal/suboptimal training structures. The proposed channel estimation strategies offer many benefits: (1) They work well when there are limited number of PTs, which saves the transmission bandwidth; (2) They extract the individual channel knowledge from the estimation and thus can help to predict the actions of the relay for data transmission, e.g., power allocation, carrier permutation, etc., without the need of feedback from the relay.

Appendix 1: Block Training Based Estimation

Proof of Theorem 4.1

From matrix inversion lemma, we know

$$tr\left((\mathbf{D}^H\mathbf{D})^{-1}\right) = tr((\mathbf{D}_1^H\mathbf{D}_1 - \mathbf{D}_1^H\mathbf{D}_2(\mathbf{D}_2^H\mathbf{D}_2)^{-1}\mathbf{D}_2^H\mathbf{D}_1)^{-1})$$

$$+ tr((\mathbf{D}_2^H\mathbf{D}_2 - \mathbf{D}_2^H\mathbf{D}_1(\mathbf{D}_1^H\mathbf{D}_1)^{-1}\mathbf{D}_1^H\mathbf{D}_2)^{-1})$$

$$= tr((\mathbf{D}_1^H P_{\mathbf{D}_2}^\perp \mathbf{D}_1)^{-1}) + tr((\mathbf{D}_2^H P_{\mathbf{D}_1}^\perp \mathbf{D}_2)^{-1}) \geq tr((\mathbf{D}_1^H\mathbf{D}_1)^{-1}) + tr((\mathbf{D}_2^H\mathbf{D}_2)^{-1}),$$

$$(4.32)$$

and the equality holds if and only if $\mathbf{D}_1^H\mathbf{D}_2 = \mathbf{0}$. However, we cannot directly claim that the optimal $\mathbf{D}_1$ and $\mathbf{D}_2$ are orthogonal since $\mathbf{D}_1$ and $\mathbf{D}_2$ have to follow the circulant structure.

Nonetheless, we first take look at the following optimization:

$$\min_{\mathbf{D}_j} tr((\mathbf{D}_j^H\mathbf{D}_j)^{-1}) \tag{4.33}$$

$$\text{s.t. } \|\mathbf{d}_j\|^2 \leq NP_j,$$

for $j = 1, 2$. Note that (4.33) is the standard optimal channel estimation in single-input single-output (SISO) OFDM [9], whose solutions are

$$|\tilde{d}_{j,i}|^2 = P_j, \quad \forall j = 1, 2, \ i = 0, \ldots, N-1. \tag{4.34}$$

Meanwhile, the minimum value of $tr((\mathbf{D}_1^H\mathbf{D}_1)^{-1})$ is $\frac{2L_1-1}{NP_1}$ and the minimum value of $tr((\mathbf{D}_2^H\mathbf{D}_2)^{-1})$ is $\frac{L_1+L_2-1}{NP_2}$.

Lastly, if we can find a pair of $\mathbf{D}_1$ and $\mathbf{D}_2$ from (4.34) that satisfies $\mathbf{D}_1^H\mathbf{D}_2 = \mathbf{0}$, then these $\mathbf{D}_1$ and $\mathbf{D}_2$ make the equality in (4.32) hold, and must be the optimal solution to (4.15).

Therefore, the training satisfying both (4.34) and the following equation

$$\sum_{i=0}^{N-1} \tilde{d}_{1,i}^* \tilde{d}_{2,i} e^{-j2\pi ki/N} = 0, \quad \forall k \in \{-(L_1+L_2-2), -(L_1+L_2-3), \ldots, (2L_1-2)\}$$

$$(4.35)$$

is the optimal solution to (4.15).

Proof of Theorem 4.2

The N-point DFT of $\mathbf{b}$ can be expressed as $\tilde{\mathbf{b}} = [\tilde{b}_0, \tilde{b}_1, \ldots, \tilde{b}_{N-1}]^T$, where $\tilde{b}_i = \tilde{h}_i^2$. Then, only $\tilde{b}_i^{1/2} = \tilde{\gamma}_i \tilde{h}_i$ can be obtained by the rooting operation, where $\tilde{\gamma}_i = \pm 1$ contains the channel uncertainty in each carrier. Define $\tilde{\boldsymbol{\gamma}} = [\tilde{\gamma}_0, \tilde{\gamma}_1, \ldots, \tilde{\gamma}_{N-1}]^T$ and $\tilde{\mathbf{t}} = [\tilde{b}_0^{1/2}, \tilde{b}_1^{1/2}, \ldots, \tilde{b}_{N-1}^{1/2}]^T = \tilde{\mathbf{h}} \odot \tilde{\boldsymbol{\gamma}}$. Construct an auxiliary vector $\tilde{\boldsymbol{\kappa}} = [\tilde{\kappa}_0, \tilde{\kappa}_1, \ldots, \tilde{\kappa}_{N-1}]^T$, where $\tilde{\kappa}_i \in \{+1, -1\}$, and define $\tilde{\boldsymbol{\zeta}} = \tilde{\boldsymbol{\gamma}} \odot \tilde{\boldsymbol{\kappa}} = [\tilde{\zeta}_0, \tilde{\zeta}_1, \ldots, \tilde{\zeta}_{N-1}]^T$, where $\tilde{\zeta}_i = \tilde{\gamma}_i \tilde{\kappa}_i$ belongs to $\{+1, -1\}$. The IDFT of $\tilde{\mathbf{t}} \odot \tilde{\boldsymbol{\kappa}}$ can then be expressed as $\mathbf{t} \otimes \boldsymbol{\kappa} = \mathbf{h} \otimes \boldsymbol{\zeta}$ where $\boldsymbol{\kappa}$ and $\boldsymbol{\zeta}$ are the IDFTs of $\tilde{\boldsymbol{\kappa}}$ and $\tilde{\boldsymbol{\zeta}}$, respectively. Suppose $\boldsymbol{\zeta}$ has length L_ζ. Then, the length of $\mathbf{h} \otimes \boldsymbol{\zeta}$ is $\min\{L_1 + L_\zeta - 1, N\}$.[6] If $L_1 = N$, then the length of $\mathbf{t}$ is N regardless of $\boldsymbol{\zeta}$. However, if $L_1 < N$, then the length of $\mathbf{h} \otimes \boldsymbol{\zeta}$ is L_1 only when $L_\zeta = 1$. In this case, $\boldsymbol{\zeta}$ has only one non-zero value at its first element and is, actually, a scale of the discrete delta function. Therefore, all $\tilde{\zeta}_i$'s are equal. Considering the range of $\tilde{\zeta}_i$, $\tilde{\boldsymbol{\zeta}}$ can only be $\pm \mathbf{1}$, which corresponds to $\tilde{\boldsymbol{\kappa}} = \pm \tilde{\boldsymbol{\gamma}}$. The so derived $\mathbf{t} \otimes \boldsymbol{\kappa}$ is then $\pm [\mathbf{h}, \mathbf{0}_{1 \times (N-L_1)}]^T$ whose first L_1 entries give $\pm \mathbf{h}$.

Proof of Theorem 4.3

After dividing the ith element from $\tilde{\mathbf{c}}$ by $\tilde{b}_i^{1/2}$, we obtain $\tilde{\mathbf{z}} = \tilde{\mathbf{c}} \oslash \tilde{\mathbf{t}} = \tilde{\mathbf{g}} \odot \tilde{\boldsymbol{\gamma}}$, where $\oslash$ denotes the element-wise division. If L_2 is known and $L_2 < N$, from a similar process in Appendix "Proof of Theorem 4.2", we can obtain $\tilde{\mathbf{g}}$ up to a sign ambiguity by restricting the length of $\mathbf{z} \otimes \boldsymbol{\kappa}$ as L_2, where $\mathbf{z}$ is the IDFT of $\tilde{\mathbf{z}}$. The effect of $\tilde{\boldsymbol{\kappa}} = \pm \tilde{\boldsymbol{\gamma}}$ can then be removed from $\tilde{\mathbf{z}}$, which results in $\pm \mathbf{g}$.

Proof of Theorem 4.4

From Appendix "Proof of Theorem 4.2", we know that non-trivial ambiguity happens if there exists a $\boldsymbol{\zeta}$ with length $L_\zeta \leq \hat{L}_1 - L_1 + 1$ such that the length of $\mathbf{t} \otimes \boldsymbol{\kappa} = \mathbf{h} \otimes \boldsymbol{\zeta}$ is still no greater than $\hat{L}_1$.

Define $\bar{\mathbf{F}}_x$ as the matrix that contains the last x columns of $\mathbf{F}$. We only need to find the maximum $\hat{L}_1$, such that $\bar{\mathbf{F}}_{N-(\hat{L}_1-L_1+1)} \tilde{\boldsymbol{\zeta}} \neq \mathbf{0}$ for all $\tilde{\boldsymbol{\zeta}}$'s except for $\tilde{\boldsymbol{\zeta}} = \pm \mathbf{1}$. We first take a look into the following lemma.

Lemma 4.1 For even N, $\bar{\mathbf{F}}_{N/2}^H \tilde{\boldsymbol{\zeta}} \neq \mathbf{0}$ for any $\tilde{\boldsymbol{\zeta}} \neq \pm \mathbf{1}$. For odd N, $\bar{\mathbf{F}}_{(N-1)/2}^H \tilde{\boldsymbol{\zeta}} \neq \mathbf{0}$ for any $\tilde{\boldsymbol{\zeta}} \neq \pm \mathbf{1}$.

---

[6]Note here, we do not consider the case where zeros appear due to the convolution between $\mathbf{h}$ and $\boldsymbol{\zeta}$ because each h_i is a continuous random variable and $\tilde{\boldsymbol{\zeta}}$ has 2^N finite states, which makes the scenario happen with probability zero.

Proof Bearing in mind that $\tilde{\boldsymbol{\zeta}}$ is a real vector, we know that for even N, if $\bar{\mathbf{F}}_{N/2}^{H}\tilde{\boldsymbol{\zeta}} = \mathbf{0}$, then there is $(\bar{\mathbf{F}}_{N/2}^{*})^{H}\tilde{\boldsymbol{\zeta}} = \mathbf{0}$. From the structure of $\mathbf{F}$, we know that the ith column of $\bar{\mathbf{F}}_{N/2}^{*}$ is the $(N/2 - i + 2)$th column of $\mathbf{F}$. Combining both $\bar{\mathbf{F}}_{N/2}^{H}\tilde{\boldsymbol{\zeta}} = \mathbf{0}$ and $(\bar{\mathbf{F}}_{N/2}^{*})^{H}\tilde{\boldsymbol{\zeta}} = \mathbf{0}$, we know $\bar{\mathbf{F}}_{N-1}^{H}\boldsymbol{\zeta} = \mathbf{0}$. Therefore, $\boldsymbol{\zeta}$ must lie in the space spanned by the first column of $\mathbf{F}$. From the range of $\tilde{\zeta}_i$, we know $\tilde{\boldsymbol{\zeta}}$ can only be ± 1.

For odd N, the ith column of $\bar{\mathbf{F}}_{(N-1)/2}^{*}$ is the $((N-1)/2 - i + 2)$th column of $\mathbf{F}$, and the rest of the proof is the same as above. $\square$

From Lemma 4.1, we know the nontrivial ambiguity does not exist for $\hat{L}_1 \leq L_1 + \frac{N}{2} - 1$. In fact, $\hat{L}_1 = L_1 + \frac{N}{2} - 1$ is also the maximum allowed value since $\bar{\mathbf{F}}_{N/2-1}^{H}\tilde{\boldsymbol{\zeta}} = \mathbf{0}$ when $\tilde{\boldsymbol{\zeta}}_i = [+1, -1, +1, -1, \ldots, +1, -1]^T$.

Noting from Theorem 4.2 that $\hat{L}_1$ should be smaller than N, we complete the proof.

Proof of Theorem 4.5

For odd N, the maximum $\hat{L}_1$ depends on the factorization of N and we do not have a general conclusion, yet. However from Lemma 4.1, we know $\hat{L}_1 \leq L_1 + \frac{N-1}{2}$ can guarantee only the sign ambiguity, although it may not be the maximum value.

Algorithm to Extract h *and* g

Based on the discussions from Appendix "Proof of Theorem 4.2" to Appendix "Proof of Theorem 4.5", $\tilde{\boldsymbol{\kappa}} = \pm \tilde{\boldsymbol{\gamma}}$ should be found from

$$\|\bar{\mathbf{F}}_{N-\hat{L}_1}^{H}(\tilde{\mathbf{t}} \odot \tilde{\boldsymbol{\kappa}})\|^2 + \|\bar{\mathbf{F}}_{N-\hat{L}_2}^{H}(\tilde{\mathbf{z}} \odot \tilde{\boldsymbol{\kappa}})\|^2 = \|\bar{\mathbf{F}}_{N-\hat{L}_1}^{H}\tilde{\mathbf{T}}\tilde{\boldsymbol{\kappa}})\|^2 + \|\bar{\mathbf{F}}_{N-\hat{L}_2}^{H}\tilde{\mathbf{Z}}\tilde{\boldsymbol{\kappa}})\|^2 = 0,$$

(4.36)

where $\tilde{\mathbf{T}} \triangleq \text{diag}\{\tilde{\mathbf{t}}\}$ and $\tilde{\mathbf{Z}} \triangleq \text{diag}\{\tilde{\mathbf{z}}\}$ are two diagonal matrices.

Due to the noise in practical transmission, the equality in (4.36) cannot hold. Then, $\tilde{\boldsymbol{\kappa}}$ should be found from

$$\tilde{\boldsymbol{\kappa}} = \arg \min_{x_i \in \{+1, -1\}} \|\boldsymbol{\Xi}\mathbf{x}\|^2,$$

(4.37)

where $\mathbf{x} = [x_0, x_1, \ldots, x_{N-1}]^T$ is the trial variable and $\boldsymbol{\Xi} \triangleq \left[(\bar{\mathbf{F}}_{N-\hat{L}_1}^{H}\tilde{\mathbf{T}})^T, (\bar{\mathbf{F}}_{N-\hat{L}_2}^{H}\tilde{\mathbf{Z}})^T\right]^T$ is a $(2N - \hat{L}_1 - \hat{L}_2) \times N$ matrix. Since the sign ambiguity always exists, we can fix $x_0 = 1$. One way is to apply the exhaustive search over 2^{N-1} possible candidates. Considering the range of x_i, we can also apply the efficient sphere decoding (SD) algorithm [12, 13] to reduce the complexity.

Appendix 2: PT Based Estimation

Proof of Theorem 4.6

Define $\tilde{\mathbf{h}}' = [\tilde{h}_{i_0}, \tilde{h}_{i_1}, \ldots, \tilde{h}_{i_{(K_1-1)}}]^T = \mathbf{F}_{K_1L_1}\mathbf{h}$, where $\mathbf{F}_{K_1L_1}$ is the matrix that contains the first L_1 columns and all the $(i_k + 1)$th rows from $\mathbf{F}$. Without loss of generality, we can assume $i_k = kQ + k_0$, where $Q = \frac{N}{K_1}$ is an integer and $0 \leq k_0 \leq Q - 1$ is a constant. In this case, $\tilde{\mathbf{h}}'$ can be equivalently considered as the K_1-point DFT of $\sqrt{\frac{N}{K_1}}\boldsymbol{\Lambda}\mathbf{h}$, where $\boldsymbol{\Lambda} = \sqrt{\frac{K_1}{N}}\mathrm{diag}\{1, e^{-j2\pi k_0/N}, \ldots, e^{-j2\pi k_0(L_1-1)/N}\}$.

Once $\tilde{\mathbf{b}}' \triangleq [\tilde{h}_{i_0}^2, \tilde{h}_{i_1}^2, \ldots, \tilde{h}_{i_{(K_1-1)}}^2]^T$ is estimated, from Theorem 4.2, we can obtain $\sqrt{\frac{N}{K_1}}\boldsymbol{\Lambda}\mathbf{h}$ with sign ambiguity, from which $\mathbf{h}$ can also be determined with sign ambiguity.

Illustration of Conjecture 4.1

For an arbitrary set $\mathcal{L}_1$, matrix $\mathbf{F}_{K_1L_1}$ has full rank because any L_1 rows from itself form a Vandermonde matrix. If $\tilde{\mathbf{h}}'$ can be obtained without ambiguity, the channel vector $\mathbf{h}$ can be recovered as $\hat{\mathbf{h}} = \mathbf{F}_{K_1L_1}^{\dagger}\tilde{\mathbf{h}}'$. Unfortunately, we can only obtain $\tilde{\mathbf{t}}' \triangleq [\tilde{\gamma}_{i_0}\tilde{h}_{i_0}, \tilde{\gamma}_{i_1}\tilde{h}_{i_1}, \ldots, \tilde{\gamma}_{i_{K_1-1}}\tilde{h}_{i_{(K_1-1)}}]^T$, where $\tilde{\gamma}_{i_k} = +1$ or $\tilde{\gamma}_{i_k} = -1$ contains the uncertainty. Define $\tilde{\boldsymbol{\gamma}}' = [\tilde{\gamma}_{i_0}, \tilde{\gamma}_{i_1}, \ldots, \tilde{\gamma}_{i_{(K_1-1)}}]^T$ and construct an auxiliary vector $\tilde{\boldsymbol{\kappa}}' = [\tilde{\kappa}_{i_0}, \tilde{\kappa}_{i_1}, \ldots, \tilde{\kappa}_{i_{(K_1-1)}}]^T$, where $\tilde{\kappa}_{i_k} \in \{+1, -1\}$. Take a look on the following possible estimator of $\mathbf{h}$:

$$\hat{\mathbf{h}} = \mathbf{F}_{K_1L_1}^{\dagger}(\tilde{\mathbf{t}}' \odot \tilde{\boldsymbol{\kappa}}') = \mathbf{F}_{K_1L_1}^{\dagger}(\tilde{\mathbf{h}}' \odot \tilde{\boldsymbol{\xi}}'), \qquad (4.38)$$

where $\tilde{\boldsymbol{\xi}}' = \tilde{\boldsymbol{\gamma}}' \odot \tilde{\boldsymbol{\kappa}}' = [\tilde{\xi}_{i_0}, \tilde{\xi}_{i_1}, \ldots, \tilde{\xi}_{i_{(K_1-1)}}]^T$ and $\tilde{\xi}_{i_k} = \tilde{\gamma}_{i_k}\tilde{\kappa}_{i_k} \in \{+1, -1\}$. If $\tilde{\boldsymbol{\xi}}' = \pm 1$, then $\hat{\mathbf{h}} = \pm\mathbf{h}$ and $\mathbf{F}_{K_1L_1}\hat{\mathbf{h}}$ is still equal to $(\tilde{\mathbf{t}}' \odot \tilde{\boldsymbol{\kappa}}')$:

$$\mathbf{F}_{K_1L_1}\mathbf{F}_{K_1L_1}^{\dagger}(\tilde{\mathbf{t}}' \odot \tilde{\boldsymbol{\kappa}}') = (\tilde{\mathbf{t}}' \odot \tilde{\boldsymbol{\kappa}}'). \qquad (4.39)$$

Note that $\mathbf{F}_{K_1L_1}\mathbf{F}_{K_1L_1}^{\dagger}$ is the projection matrix on the space spanned by $\mathbf{F}_{K_1L_1}$. For (4.39) to hold, we need $(\tilde{\mathbf{t}}' \odot \tilde{\boldsymbol{\kappa}}')$ to stay in the span of $\mathbf{F}_{K_1L_1}$. Equation (4.39) obviously holds if $\tilde{\boldsymbol{\xi}}' = \pm 1$.

Intuitively, since channels are continuous random variable, $\tilde{\mathbf{h}}'$, who originally stays in the space spanned by $\mathbf{F}_{K_1 L_1}$ in general, does not stay in the same space after taking inner product with $\tilde{\boldsymbol{\zeta}}' \neq \pm \mathbf{1}$. Moreover, we must require $K_1 > L_1$ since (4.39) always holds for $K_1 = L_1$, which gives $\mathbf{F}_{K_1 L_1} \mathbf{F}_{K_1 L_1}^\dagger = \mathbf{I}_{K_1}$.

Proof of Theorem 4.7

Once $K_1 > L_1$, we can find $\hat{\mathbf{h}} = \pm \mathbf{h}$. Then, $\tilde{h}_{q_k}$, $q_k \in \mathcal{L}_2$ can be determined within sign ambiguity from $\mathbf{F}_{K_2 L_1} \hat{\mathbf{h}}$, where $\mathbf{F}_{K_2 L_1}$ is the matrix that contains all $q_k \in \mathcal{L}_2$ rows and the first L_1 columns of $\mathbf{F}$. Then, $\tilde{g}_{q_k}$ can be determined with SSA from $\tilde{\mathbf{c}}' \oslash (\mathbf{F}_{K_2 L_1} \hat{\mathbf{h}})$. Finally, $\hat{\mathbf{g}} = \mathbf{F}_{K_2 L_2}^\dagger (\tilde{\mathbf{c}}' \oslash (\mathbf{F}_{K_2 L_1} \hat{\mathbf{h}})) = \pm \mathbf{g}$.

Efficient Algorithm to Estimate h and g

Based on the discussions in Appendix "Proof of Theorem 4.6" and Appendix "Illustration of Conjecture 4.1", $\tilde{\boldsymbol{\kappa}}'$ should be found from

$$\|(\mathbf{I} - \mathbf{F}_{K_1 L_1} \mathbf{F}_{K_1 L_1}^\dagger)(\tilde{\mathbf{t}}' \odot \tilde{\boldsymbol{\kappa}}')\|^2 = \|(\mathbf{I} - \mathbf{F}_{K_1 L_1} \mathbf{F}_{K_1 L_1}^\dagger)\tilde{\mathbf{T}}' \tilde{\boldsymbol{\kappa}}'\|^2 = 0 \qquad (4.40)$$

where $\tilde{\mathbf{T}}'$ is the diagonal matrix with the diagonal elements $\tilde{\mathbf{t}}'$. When K_1 PTs from $\mathcal{L}_1$ are equispaced, it can be verified that (4.40) is equivalent to checking whether the length of $\mathbf{F}_{K_1 L_1}^\dagger (\tilde{\mathbf{t}}' \odot \tilde{\boldsymbol{\kappa}}')$ is L_1. Note that, $\tilde{\mathbf{c}}'$ does not contribute to the determination of $\tilde{\boldsymbol{\kappa}}'$ here.

In practical transmission, $\tilde{\boldsymbol{\kappa}}'$ should be found from

$$\tilde{\boldsymbol{\kappa}}' = \arg \min_{x_i \in \{+1, -1\}} \|(\mathbf{I} - \mathbf{F}_{K_1 L_1} \mathbf{F}_{K_1 L_1}^\dagger)\tilde{\mathbf{T}}' \mathbf{x}\|^2. \qquad (4.41)$$

Again, the SD algorithm [12, 13] can be carried out to reduce the computational complexity.

References

1. T. Cui, F. Gao, T. Ho, and A. Nallanathan. Distributed space-time coding for two-way wireless relay networks. in *Proc. of IEEE ICC*, Beijing, China, May 2008, pp. 3888–3892.
2. T. Cui, T. Ho, and J. Kliewer. Memoryless relay strategies for two-way relay channels: performance analysis and optimization. in *Proc. of IEEE ICC*, pp. 1139–1143, Beijing, China, May 2008.

3. R. Zhang, Y.-C. Liang, C. C. Chai, and S. G. Cui. Optimal beamforming for two-way multi-antenna relay channel with analogue network coding. *IEEE J. Select. Areas in Commun.*, 27(5),pp. 699–712 , June 2009.

4. S. Ohno and G. B. Giannakis. Optimal training and redundant precoding for block transmissions with application to wireless OFDM. *IEEE Trans. Commun.*, 50(12), pp. 2113–2123, Dec. 2002.

5. I. Barhumi, G. Leus, and M. Moonen. Optimal trianing design for MIMO OFDM systems in mobile wireless channels. *IEEE Trans. Signal Processing*, 51(5), pp. 1615–1624, June 2003.

6. H. Minn and N. Al-Dhahir. Optimal training signals for MIMO OFDM channel estimation. *IEEE Trans. Wireless Commun.*,5(6), pp. 158–1168, May 2006.

7. C. K. Ho, R. Zhang, and Y.-C. Liang. Two-way relaying over OFDM: optimized tone permutation and power allocation. in *Proc. of IEEE ICC*, pp. 3908–3912, Beijing China, May 2008.

8. "Wireless LAN medium access control (MAC) and physical layer (PHY) specifications: high speed physical layer in the 5 GHZ band," IEEE802.11a, 1999.

9. J. H. Manton. Optimal trianing sequences and pilot tone for OFDM systems. *IEEE Commun. Lett.*, 5(4), pp. 151–153, Apr. 2001.

10. S. Boyd and L. Vandenberghe, *Convex Optimization*,   Cambridge University Press, 2006.

11. F. Gao and A. Nallanathan. Blind channel estimation for MIMO OFDM systems via non-redundant linear precoding. *IEEE Trans. Signal Processing*, 55(2), pp. 784–789, Jan. 2007.

12. B. Hassibi and H. Vikalo. On the sphere decoding algorithm I: expected complexity. *IEEE Trans. Signal Processing*, 53(8), pp. 2806–2818, Aug. 2005.

13. T. Cui and C. Tellambura. An efficient generalized sphere decoder for rank-deficient MIMO systems. *IEEE Commun. Lett.*, 9(5), pp. 423–425, May 2005.

Chapter 5
Channel Estimation for PLNC Under Time-Selective Fading Scenario

Abstract In this chapter, we design channel estimation and training sequence for PLNC in a time-selective fading environment. We first propose a new complex-exponential basis expansion model (CE-BEM) to represent the mobile-to-mobile time-varying channels. To estimate such channels, a novel pilot symbol-aided transmission scheme is developed such that a low complex linear approach can be applied to estimate the BEM coefficients of the convoluted channels. More essentially, we design two algorithms to extract the BEM coefficients of the individual channels. The optimal training parameters, including the number of the pilot symbols, the placement of the pilot symbols, and the power allocation to the pilot symbols, are derived by minimizing the channel mean-square error (MSE). Finally, extensive numerical results are provided to corroborate the proposed studies.

5.1 System Model

Consider a PLNC with two source nodes $\mathbb{T}_1$, $\mathbb{T}_2$ and one relay node $\mathbb{R}$, as shown in Fig. 5.1. Each node has only one antenna and operates in the half-duplex mode. The baseband channel from $\mathbb{T}_i$, $i = 1, 2$ to $\mathbb{R}$ is assumed to be time-selective flat-fading and is denoted by $h_i(n)$, where n is the discrete time index. Moreover, the channels are modeled as wide-sense stationary (WSS) zero mean complex Gaussian random processes with variances $\sigma_{h_i}^2$. With TDD mode, the channel from $\mathbb{R}$ to $\mathbb{T}_i$ is also denoted as $h_i(n)$.

5.1.1 Time Varying Relay Channels

The channel statistics in a relay network depend on the mobility of the three nodes [1]. Denote f_{d1}, f_{d2} and f_{dr} as the maximum Doppler shifts due to the motion of $\mathbb{T}_1$, $\mathbb{T}_2$, and $\mathbb{R}$, respectively. The discrete autocorrelation functions of $h_i(n)$'s can be represented as [2]

© The Author(s) 2014
F. Gao et al., *Channel Estimation for Physical Layer Network Coding Systems*,
SpringerBriefs in Computer Science, DOI 10.1007/978-3-319-11668-6_5

$$\text{T}_1 \xleftrightarrow{\quad h_1(n) \quad} \text{R} \xleftrightarrow{\quad h_2(n) \quad} \text{T}_2$$

Fig. 5.1 A PLNC system over time-selective flat-fading channels

$$R_{h_i}(m) = \text{E}\{h_i(n+m)h_i^*(n)\} = \sigma_{h_i}^2 J_0(2\pi f_{di} m T_s) J_0(2\pi f_{dr} m T_s), \quad i = 1, 2 \tag{5.1}$$

where $J_0(\cdot)$ is the zero-th order Bessel function of the first kind, and T_s is the symbol sampling time. The correlation function in (5.1) has been widely adopted to describe the mobile-to-mobile link [1, 2]. If one node is fixed, i.e., if the corresponding Doppler shift is zero, then (5.1) reduces to the well-known Jakes model [3]. Meanwhile, (5.1) reveals that the power spectra of $h_1(n)$ and $h_2(n)$ span over the bandwidths $f_1 = f_{d1} + f_{dr}$ and $f_2 = f_{d2} + f_{dr}$, respectively, which indicates an increased Doppler effect from the mobile-to-mobile transmission.

We apply the parsimonious finite-parameter BEM [4] to approximate the two time-varying channels to reduce the number of the parameters, i.e., during any time interval of NT_s, $h_i(n)$'s can be approximated by

$$h_1(n) = \sum_{q=0}^{Q_1} \lambda_q w_1(q), \qquad h_2(n) = \sum_{q=0}^{Q_2} \mu_q w_2(q), \quad 0 \le n \le N - 1, \tag{5.2}$$

where λ_q's and μ_q's are the BEM coefficients that remain invariant within one interval but will change in the next interval, while $w_i(q)$'s are the bases that will remain unchanged for any interval. The number of the bases Q_i is a function of the channel bandwidth f_i and the interval length NT_s. Specific choices for $\{w_i(q)\}_{q=0}^{Q_i}$ include the polynomial [5], wavelet [6], discrete prolate spheroid [7], and Fourier bases [8]. In this chapter, we choose the CE-BEM [9] that is a specific form of Fourier bases. Then (5.2) can be explicitly written as

$$h_1(n) = \sum_{q=0}^{Q_1} \lambda_q e^{j 2\pi (q - Q_1/2) n/N}, \quad 0 \le n \le N - 1, \tag{5.3a}$$

$$h_2(n) = \sum_{q=0}^{Q_2} \mu_q e^{j 2\pi (q - Q_2/2) n/N}, \quad 0 \le n \le N - 1. \tag{5.3b}$$

CE-BEM (5.3) can be viewed as the Fourier series of the time-varying channels, and the number of bases Q_i should be at least $2\lceil f_i N T_s \rceil$ in order to provide sufficient degrees of freedom [8,9]. Moreover, the larger the Q is, the better the approximation will be.

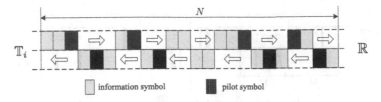

Fig. 5.2 Proposed transmission strategy for PLNC with time-varying channel

To simplify the notation as well as the discussion, we assume $f_1 = f_2 = f_d$ and $Q_1 = Q_2 = Q$. Nonetheless, the extension to the general cases is straightforward. We further denote $\omega_q = 2\pi(q - Q/2)/N$ and define

$$\boldsymbol{\lambda} = [\lambda_0, \lambda_1, \ldots, \lambda_Q]^T, \qquad \boldsymbol{\mu} = [\mu_0, \mu_1, \ldots, \mu_Q]^T$$

for subsequent use.

5.1.2 Transmission Strategy

We propose a new transmission strategy over one interval NT_s for PLNC, as depicted in Fig. 5.2. Let $\mathcal{D}_t$ and $\mathcal{T}_t$ be the time index sets for the transmitted information symbols and the pilot symbols from $\mathbb{T}_i$, $i = 1, 2$, respectively. Moreover, let $\mathcal{D}_r$ and $\mathcal{T}_r$ be the time index sets for the received information symbols and pilot symbols at $\mathbb{T}_i$, respectively. These four sets are disjoint with the property that $\mathcal{D}_t \bigcup \mathcal{T}_t \bigcup \mathcal{D}_r \bigcup \mathcal{T}_r = \{0, 1, \ldots, N - 1\}$. Let us define the cardinality of the sets as $|\mathcal{D}_t| = |\mathcal{D}_r| = D$ and $|\mathcal{T}_t| = |\mathcal{T}_r| = T$. Then, $N = 2(D + T)$ is an even integer.

We assume that the relay node $\mathbb{R}$ forwards its received symbols of time slot $g(n)$ to both $\mathbb{T}_i$ on time slot n; i.e.,

$$\mathcal{D}_t \bigcup \mathcal{T}_t = \left\{ g(n) | n \in \mathcal{D}_r \bigcup \mathcal{T}_r \right\}. \tag{5.4}$$

It is possible to optimize $g(n)$ according to different criteria, i.e., data detection MSE, bit-error-rate (BER), throughput, and others. Note that $0 \leq g(n) < n$ is required because $\mathbb{R}$ can only forward a symbol after receiving it. Interestingly, the conventional data transmission in a PLNC becomes a special case of our proposed scheme if $g(n) = n - N/2$ is selected.

Remark. A special yet important case involves evenly dividing NT_s intervals into several sub-blocks, as shown in Fig. 5.3. This case corresponds to setting $g(n) = n - M$, where M divides $N/2$, and will be separately discussed later. Whether to adopt the general scheme (Fig. 5.2) or the sub-block-based scheme (Fig. 5.3) depends on the synchronization requirement in practical scenarios and other design issues.

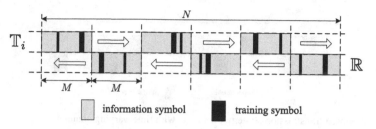

Fig. 5.3 Sub-block based transmission strategy

Denote the symbols sent from $\mathbb{T}_i$ as $s_i(n)$, $n \in \mathcal{D}_t \bigcup \mathcal{T}_t$, of which the average power for the information symbols is P_i; i.e., $\mathrm{E}\{|s_i(n)|^2\} = P_i$, $\forall n \in \mathcal{D}_t$, while the total training power[1] is $P_{i,t}$; i.e., $\sum_{n \in \mathcal{T}_t} |s_i(n)|^2 = P_{i,t}$. With perfect synchronization, $\mathbb{R}$ receives

$$r(n) = h_1(n)s_1(n) + h_2(n)s_2(n) + w_r(n), \quad n \in \mathcal{D}_t \bigcup \mathcal{T}_t, \tag{5.5}$$

where $w_r(n)$ is the circularly symmetric complex Gaussian noise with the variance σ_r^2. If the average transmit power of $\mathbb{R}$ is P_r, then $r(n)$ will be scaled by

$$\alpha(n) = \begin{cases} \sqrt{\dfrac{P_r}{\sigma_{h_1}^2 P_1 + \sigma_{h_2}^2 P_2 + \sigma_r^2}} & n \in \mathcal{D}_r \\[2ex] \sqrt{\dfrac{P_r}{\sigma_{h_1}^2 P_{1,t}/T + \sigma_{h_2}^2 P_{2,t}/T + \sigma_r^2}} & n \in \mathcal{T}_r \end{cases} \tag{5.6}$$

before it is forwarded to $\mathbb{T}_i$'s to keep the power constraint.

Remark. More practical considerations should include the process delay at $\mathbb{R}$ as well as the path-delay between $\mathbb{T}_1$ and $\mathbb{T}_2$. These considerations results in only slightly changing the channel from $h_i(n)$ to $h_i(n + \Delta n)$, and the remaining discussion holds the same.

5.1.3 Channel Estimation Strategy

Due to symmetry, we only present the estimation procedure at $\mathbb{T}_1$, and the received signal is

---

[1] We should not consider the average power constraints for training because otherwise, the training length is trivially preferred to be as large as possible.

$$y(n) = \alpha(n)h_1(n)r(g(n)) + w_1(n)$$

$$= \alpha(n)\underbrace{h_1(n)h_1(g(n))}_{b_1(n)}s_1(g(n)) + \alpha(n)\underbrace{h_1(n)h_2(g(n))}_{b_2(n)}s_2(g(n))$$

$$+ \underbrace{\alpha(n)h_1(n)w_r(g(n)) + w_1(n)}_{w(n)}, \qquad n \in \mathcal{D}_r \bigcup \mathcal{T}_r, \qquad (5.7)$$

where $w_1(n)$ is the complex circularly symmetric Gaussian noise at $\mathbb{T}_1$ with variance σ_1^2; $w(n)$ denotes the overall noise; and $b_i(n)$, $i = 1, 2$ can be treated as the equivalent time-varying channel of $\mathbb{T}_i \to \mathbb{R} \to \mathbb{T}_1$. Obviously, if $b_i(n)$'s are known at $\mathbb{T}_1$, the self-signal component $s_1(g(n))$ can be subtracted from $y(n)$ in order to detect the desired information $s_2(g(n))$.

To gain more insight into the time-varying channels, we apply (5.3) and rewrite $b_i(n)$'s as

$$b_1(n) = \sum_{p=0}^{Q}\sum_{q=0}^{Q}\lambda_p\lambda_q e^{j(\omega_p n + \omega_q g(n))}, \quad b_2(n) = \sum_{p=0}^{Q}\sum_{q=0}^{Q}\lambda_p\mu_q e^{j(\omega_p n + \omega_q g(n))},$$

$$n \in \mathcal{D}_r \bigcup \mathcal{T}_r. \qquad (5.8)$$

The new expression (5.8) indicates that in order to obtain $b_i(n)$, $0 \le n \le N - 1$, one needs to know either $2(Q + 1)$ parameters $\lambda_p, \mu_p, p = 0, \ldots, Q$ or $2(Q + 1)^2$ parameters $\lambda_p\lambda_q, \lambda_p\mu_q, p, q = 0, \ldots, Q$. For a general mapping function $g(n)$, the former approach requires a non-linear search, which is computationally prohibitive, while the latter approach, though could be implemented from linear approach, possesses large redundancy in the number of estimated variables.

To facilitate the channel estimation, we propose to use

$$g(n) = n - M, \qquad (5.9)$$

for $n \in \mathcal{T}_r$, while $g(n)$ for information transmission $n \in \mathcal{D}_r$ could still be designed from certain optimization criterion. The condition (5.9) says that $\mathbb{R}$ retransmits each received pilot symbol with a delay of M-symbol interval, and this interval is common for all pilot symbols.

With (5.9), the received pilot symbols at $\mathbb{T}_1$ can be further expressed as

$$y(n) = \alpha \sum_{m=0}^{2Q}\underbrace{\left(\sum_{q=0}^{m}\lambda_{m-q}\lambda_q e^{-j\omega_q M}\right)}_{x_1(m)}e^{j\theta_m n}s_1(n - M)$$

$$+ \alpha \sum_{m=0}^{2Q}\underbrace{\left(\sum_{q=0}^{m}\lambda_{m-q}\mu_q e^{-j\omega_q M}\right)}_{x_2(m)}e^{j\theta_m n}s_2(n - M) + w(n), \quad n \in \mathcal{T}_r,$$

$$(5.10)$$

where $\theta_m = 2\pi(m - Q)/N$, $x_i(m)$ are defined as the corresponding items, and the index n in $\alpha(n)$ is omitted for brevity. Note that when deriving (5.10), we use the property that $\omega_p + \omega_q = \omega_{p'} + \omega_{q'}$ whenever $p + q = p' + q'$.

Remark. If the sub-block transmission in Fig. 5.3 is applied, then (5.10) is also applicable for the received information symbols $n \in \mathcal{D}_r$.

Interestingly, we may treat $x_i(m)$'s as the equivalent BEM coefficients with $2Q+1$ carriers $e^{j\theta_m n}$ to represent the equivalent time-varying channel $b_i(n)$, $n \in \mathcal{T}_r$. Moreover, the equivalent BEM sequence $x_1(m)$ is the convolution between original BEM λ_p and $e^{-j\omega_q M}\lambda_q$, while $x_2(m)$ is the convolution between λ_p and $e^{-j\omega_p M}\mu_q$.

Define

$$\mathbf{x}_i = [x_i(0), x_i(1), \ldots, x_i(2Q)]^T, \qquad \boldsymbol{\Gamma} = \mathrm{diag}\{e^{-j\omega_0 M}, e^{-j\omega_1 M}, e^{-j\omega_Q M}\}$$

and define $\boldsymbol{\Lambda}$ as the $(2Q + 1) \times (Q + 1)$ Toeplitz matrix with the first column $[\boldsymbol{\lambda}^T, \mathbf{0}_{1 \times Q}]^T$. We can explicitly express the convolutions as

$$\mathbf{x}_1 = \boldsymbol{\lambda} \otimes (\boldsymbol{\Gamma}\boldsymbol{\lambda}) = \boldsymbol{\Lambda}\boldsymbol{\Gamma}\boldsymbol{\lambda}, \qquad \text{and} \qquad \mathbf{x}_2 = \boldsymbol{\lambda} \otimes (\boldsymbol{\Gamma}\boldsymbol{\mu}) = \boldsymbol{\Lambda}\boldsymbol{\Gamma}\boldsymbol{\mu}. \tag{5.11}$$

Based on (5.10), we may estimate the equivalent BEM coefficient $x_i(m)$ (with $4Q + 2$ unknowns) and recover the original BEM λ_q, μ_q (with $2Q + 2$ unknowns). Then, the equivalent time-varying channels $b_i(n)$, $n \in \mathcal{D}_r$ can be obtained from (5.8).

5.2 Channel Estimation and Training Design

Let us specify the indices in $\mathcal{T}_r$ as $n_0 < n_1 \ldots < n_{T-1}$, and define

$$\mathbf{y}_t = [y(n_0), y(n_1), \ldots, y(n_{T-1})]^T, \qquad\qquad \mathbf{w}_t = [w(n_0), w(n_1), \ldots, w(n_{T-1})]^T,$$
$$\mathbf{t}_i = [s_i(n_0 - M), s_i(n_1 - M), \ldots, s_i(n_{T-1} - M)]^T, \quad \mathbf{T}_i = \mathrm{diag}\{\mathbf{t}_i\}, \quad i = 1, 2,$$

where $\mathbf{t}_i$ contains all the pilot symbols from $\mathbb{T}_i$. For notational simplicity, the mth entry of $\mathbf{t}_i$ is also denoted by $t_i(m)$, $m = 0, \ldots, T - 1$.

With the aid of (5.10), we can express $\mathbf{y}_t$ in matrix form as

$$\mathbf{y}_t = \alpha\mathbf{T}_1\mathbf{A}\mathbf{x}_1 + \alpha\mathbf{T}_2\mathbf{A}\mathbf{x}_2 + \mathbf{w}_t, \tag{5.12}$$

where $\mathbf{A}$ is the $T \times (2Q + 1)$ matrix

$$\mathbf{A} = \begin{bmatrix} e^{j\theta_0 n_0} & e^{j\theta_1 n_0} & \cdots & e^{j\theta_{2Q} n_0} \\ e^{j\theta_0 n_1} & e^{j\theta_1 n_1} & \cdots & e^{j\theta_{2Q} n_1} \\ \vdots & \vdots & \cdots & \vdots \\ e^{j\theta_0 n_{T-1}} & e^{j\theta_1 n_{T-1}} & \cdots & e^{j\theta_{2Q} n_{T-1}} \end{bmatrix}. \tag{5.13}$$

5.2.1 Channel Estimation

When $T \geq 4Q + 2$, there are sufficient observations to estimate all the unknown $x_i(m)$'s. In this case, one could choose a linear estimator, e.g., LS or LMMSE estimator, to reduce the computational complexity. We here choose LS estimator to present the estimation approach while the LMMSE estimator can be similarly design when the statistics of the BEM coefficients are available [8, 10, 11]. Nevertheless, LS estimator performs similarly to LMMSE estimator at relatively high SNR.

Let us define

$$\mathbf{T} = [\mathbf{T}_1 \mathbf{A}, \mathbf{T}_2 \mathbf{A}], \qquad \mathbf{x} = [\mathbf{x}_1^T, \mathbf{x}_2^T]^T.$$

The LS estimator of $\mathbf{x}$ is expressed as

$$\hat{\mathbf{x}} = \frac{1}{\alpha} \mathbf{T}^\dagger \mathbf{y} = \frac{1}{\alpha} (\mathbf{T}^H \mathbf{T})^{-1} \mathbf{T}^H \mathbf{y}, \tag{5.14}$$

with the error covariance matrix given by

$$\mathbf{W} = \mathbf{T}^\dagger \begin{bmatrix} \sigma_r^2 |h_1(n_0)|^2 + \frac{\sigma_i^2}{\alpha^2} & \cdots & 0 \\ \vdots & \ddots & \vdots \\ 0 & \cdots & \sigma_r^2 |h_1(n_{T-1})|^2 + \frac{\sigma_i^2}{\alpha^2} \end{bmatrix} (\mathbf{T}^\dagger)^H. \tag{5.15}$$

5.2.2 Training Sequence Design

The channel estimation MSE is defined as $\mathrm{tr}(\mathbf{W})$ and is related to the instant CSI that cannot be directly worked with. Hence, we propose to minimize the average MSE, defined as

$$\mathrm{AMSE} = \mathrm{E}_h\{\mathrm{tr}(\mathbf{W})\} = \left(\sigma_{h_1}^2 \sigma_r^2 + \frac{\sigma_i^2}{\alpha^2} \right) \mathrm{tr}((\mathbf{T}^H \mathbf{T})^{-1}), \tag{5.16}$$

where the property $J_0(0) = 1$ is used. We further partition $(\mathbf{T}^H \mathbf{T})^{-1}$ as

$$(\mathbf{T}^H \mathbf{T})^{-1} = \begin{bmatrix} \mathbf{A}^H \mathbf{T}_1^H \mathbf{T}_1 \mathbf{A} & \mathbf{A}^H \mathbf{T}_1^H \mathbf{T}_2 \mathbf{A} \\ \mathbf{A}^H \mathbf{T}_2^H \mathbf{T}_1 \mathbf{A} & \mathbf{A}^H \mathbf{T}_2^H \mathbf{T}_2 \mathbf{A} \end{bmatrix}^{-1}. \qquad (5.17)$$

The optimal training design amounts to selecting the number of the pilot symbols, their placement, and the power allocation for each pilot by minimizing the AMSE. The optimization problem is then formulated as

$$\text{(P1):} \qquad \min_{t_1, t_2, \mathcal{T}_r} \quad \left(\sigma_{h_1}^2 \sigma_r^2 + \frac{\sigma_1^2}{\alpha^2} \right) \text{tr}((\mathbf{T}^H \mathbf{T})^{-1}) \qquad (5.18)$$

$$\text{s.t.} \quad \sum_{m=0}^{T-1} |t_i(m)|^2 \le P_{i,t}, \quad i = 1, 2.$$

Since α is related only to T, we can first solve the following problem, for a given T:

$$\text{(P2):} \qquad \min_{\substack{t_1, t_2, \\ n_i : 0 \le i \le T}} \quad \text{tr}((\mathbf{T}^H \mathbf{T})^{-1}) \qquad (5.19)$$

$$\text{s.t.} \quad \sum_{m=0}^{T-1} |t_i(m)|^2 \le P_{i,t}, \quad i = 1, 2.$$

From [12], we know that

$$\text{tr}((\mathbf{T}^H \mathbf{T})^{-1}) \ge \sum_{i=0}^{4Q+1} \frac{1}{[\mathbf{T}^H \mathbf{T}]_{i,i}} = \sum_{i=1}^{2} \frac{2Q+1}{\sum_{m=0}^{T-1} |t_i(m)|^2}, \qquad (5.20)$$

where $[\mathbf{T}^H \mathbf{T}]_{i,i}$ is the ith diagonal elements of $\mathbf{T}^H \mathbf{T}$, and the equality holds when $\mathbf{T}^H \mathbf{T}$ is a diagonal matrix. Let us first formulate a new optimization problem:

$$\text{(P3):} \qquad \min_{t_1, t_2} \quad \sum_{i=1}^{2} \frac{2Q+1}{\sum_{m=0}^{T-1} |t_i(m)|^2} \qquad (5.21)$$

$$\text{s.t.} \quad \sum_{m=0}^{T-1} |t_i(m)|^2 \le P_{i,t}, \quad i = 1, 2.$$

Obviously, the optimal objective of (P3) serves as a lower bound for (P2). Since (P3) is a simple convex optimization, any training sequence satisfying $\sum_{m=0}^{T-1} |t_i(m)|^2 = P_{i,t}$ is optimal. Hence, if we can find t_i's that satisfy the equality constraints and make $\mathbf{T}^H \mathbf{T}$ diagonal, then these t_i's must also be the optimal solutions for problem (P2). In other words, the sufficient conditions for the optimal solutions to (P2) are

$$\mathbf{A}^H \mathbf{T}_i^H \mathbf{T}_i \mathbf{A} = P_{i,t} \mathbf{I}_{2Q+1}, \quad i = 1, 2, \tag{5.22a}$$

$$\mathbf{A}^H \mathbf{T}_1^H \mathbf{T}_2 \mathbf{A} = \mathbf{0}_{2Q+1}. \tag{5.22b}$$

Observing the Vandermonde structure of $\mathbf{A}$ and the structure of θ_m, we know that if the pilot symbols are equi-powered and equi-spaced over $\{0, \ldots, N-1\}$, then (5.22a) is satisfied; i.e.,

(C1) : $\qquad |t_i(m)|^2 = P_{i,t}/T, \quad \forall m = 0, 1, \ldots, T-1, \quad i = 1, 2,$

(C2) : $\qquad n_m = mL + l_0, \qquad \forall l_0 \in [M, L-1], \quad$ and $L = N/T$ is an integer,

where we include the consideration that $n_0 \geq M$ in (C2).[2] Combined with (C1) and (C2), the following condition can guarantee (5.22b):

$$\sum_{m=0}^{T-1} t_2^*(m) t_1(m) e^{-j\theta_u n_m} e^{-j\theta_v n_m} = 0, \quad \forall u, v = 0, 1, \cdots, 2Q$$

which can be simplified as

(C3) : $\qquad \displaystyle\sum_{m=0}^{T-1} t_2^*(m) t_1(m) e^{j 2\pi mk/T} = 0, \quad \forall k = -2Q, -2Q+1, \ldots, 2Q.$

One example of pilot sequences that satisfy conditions (C1)–(C3) is

$$\mathbf{t}_1 = \sqrt{\frac{P_{1,t}}{T}} [+1, +1, +1, \ldots, +1, +1]^T, \tag{5.23a}$$

$$\mathbf{t}_2 = \sqrt{\frac{P_{2,t}}{T}} [1, e^{j 2\pi v/T}, \ldots, e^{j 2\pi (T-1)v/T}]^T, \quad \forall v = 2Q+1, \ldots, T-2Q-1. \tag{5.23b}$$

The minimum $\mathrm{tr}((\mathbf{T}^H \mathbf{T})^{-1})$ is then $(2Q+1)(1/P_{1,t} + 1/P_{2,t})$ and does not depend on T. Hence, the optimal value of T should be independently obtained from

$$T = \arg\min_T \left(\sigma_{h_1}^2 \sigma_r^2 + \frac{\sigma_1^2}{\alpha^2} \right) = \arg\max_T \frac{P_r}{\sigma_{h_1}^2 P_{1,t}/T + \sigma_{h_2}^2 P_{2,t}/T + \sigma_r^2}. \tag{5.24}$$

The objective function (5.24) is an increasing function of T, so the optimal T should be made as large as possible. Note that this result is different from the conventional

---

[2] $n_0 = l_0$ denotes the index of the first symbol sent by $\mathbb{R}$. From the adopted $g(n)$, $n_0 = l_0 \geq M$ is required.

training design in point-to-point systems, where the channel estimation MSE is related only with the total training power but not to the training length.

However, increasing T would reduce the efficiency of the data transmission and, consequently, the system throughput. Besides, the constant $\sigma_{h_1}^2 \sigma_r^2$ will dominate the summation from $\left(\sigma_{h_1}^2 \sigma_r^2 + \frac{\sigma_1^2}{\alpha^2}\right)$ when T is greater than a certain threshold. Therefore, increasing T beyond a certain value cannot improve the channel estimation MSE, but the throughput will be linearly decreased. We here simply consider achieving the minimum amount of training as our optimization goal, while a more meaningful design of T can be obtained by maximizing the transmission throughput criterion [8, 10, 11].

The selection of the minimum possible T depends on many factors and will be discussed in the next subsection. When $T = 4Q + 2$ is allowed, the optimal pilot schemes become more specific:

$$\mathbf{t}_1 = \sqrt{\frac{P_{1,t}}{4Q+2}}[+1, +1, +1, +1, \ldots, +1, +1]^T,$$

$$\mathbf{t}_2 = \sqrt{\frac{P_{2,t}}{4Q+2}}[+1, -1, +1, -1, \ldots, +1, -1]^T,$$

and the corresponding minimum AMSE is

$$\text{AMSE} = \left((2Q+1)\sigma_{h_1}^2 \sigma_r^2 + \frac{\left(\sigma_{h_1}^2 P_{1,t} + \sigma_{h_2}^2 P_{2,t} + 2\sigma_r^2(2Q+1)\right)\sigma_1^2}{2P_r}\right)\left(\frac{1}{P_{1,t}} + \frac{1}{P_{2,t}}\right).$$
(5.25)

Remark. Importantly, it can be verified that the designed optimal pilot sequences for channel estimation at $\mathbb{T}_1$ are also optimal at $\mathbb{T}_2$. Hence, simultaneous optimal channel estimation can be achieved at both source nodes.

5.2.3   Parameter Selection

Observing (C2), we know the following: (i) The pilot spacing L should at least be $M + 1$; (ii) To transmit non-zero information symbols in one NT_s interval, we need $D = \frac{LT - 2T}{2} \geq 1$, so the spacing L should be at least 3;[3] (iii) Since N must be even, either T or L should be an even integer.

---

[3]This conclusion is also seen from the fact that if $L = 2$, then the only choice for M is 1, in which case $\mathbb{T}_i$ alternatively transmits and receives pilot symbols while no information can be sent.

The above discussion suggests the guidelines for choosing T, i.e., select the smallest integer that is greater than or equal to $4Q + 2$, divides N, and satisfies $N/T \geq 3$.

Remark. Since $T \geq 4Q + 2$ pilot symbols are needed to provide sufficient observations, and since $Q \geq 1$ for a time-varying channel, the PLNC requires that pilot symbols be transmitted back-and-forth at least six times. Therefore, the conventional PLNC frame transmission structure, i.e., sending and receiving the continuous data sequence only once, obviously does not work in time-selective environment.

Remark. For the sub-block based frame structure in Fig. 5.3, the receiving equi-spaced pilot at $\mathbb{T}_i$ is possible only if each sub-block contains one pilot symbol at the same position of each sub-block.

5.2.4 Extracting the Single Hop BEM Coefficient

After estimating $\mathbf{x}_i$'s, $i = 1, 2$, we need to obtain the original BEM coefficients λ_q and μ_q in order to build the time-varying channel $b_i(n)$, $n \in \mathcal{D}_r$. Retrieving λ_q and μ_q from $\mathbf{x}_i$ generally requires solving multivariate nonlinear equations, but doing so is computationally quite expensive. In the following, we propose two simple methods which are described under a noise-free scenario.

1. *Time-Domain Approach*: Because of the structure of $x_1(m)$, a straightforward way is to estimate λ_q sequentially. Specifically, we first estimate λ_0 from

$$\lambda_0 = I_s \left(x_1(0) e^{j\omega_0 M} \right)^{1/2}, \tag{5.26}$$

where $I_s = \pm 1$ denotes the sign uncertainty. By choosing any of the positive or negative signs in (5.26), λ_1 can be computed from

$$\lambda_1 = \frac{x_1(1)}{\lambda_0 e^{-j\omega_0 M} + \lambda_0 e^{-j\omega_1 M}}. \tag{5.27}$$

We then sequentially compute λ_q from $x_1(q)$ with the previous estimates of $\lambda_0, \ldots, \lambda_{q-1}$. The detailed steps are straightforward and are omitted here.

The above process uses only the first $Q + 1$ entries in $\mathbf{x}_1$ and cannot provide satisfactory precision. Nevertheless, with this initial estimation, we can apply the gradient decent process [13] to improve the estimation accuracy. The objective is to minimize the distance between $\mathbf{x}$ and $\Lambda \Gamma \lambda$; i.e., $\zeta = \|\mathbf{x} - \Lambda \Gamma \lambda\|^2$. Then, λ can be updated according to

$$\lambda^{(i+1)} = \lambda^{(i)} - \epsilon \left. \frac{\partial \zeta}{\partial \lambda^*} \right|_{\lambda = \lambda^{(i)}}, \tag{5.28}$$

where ϵ is the step size. The partial differential in (5.28) can be explicitly expressed as

$$\frac{\partial \zeta}{\partial \lambda^*} = -(\Lambda \Gamma + \Omega)^H (x - \Lambda \Gamma \lambda), \tag{5.29}$$

where Ω is a $(2Q + 1) \times (Q + 1)$ Toeplitz matrix with the first column $[(\Gamma \lambda)^T, \mathbf{0}_{1 \times Q}]^T$.

Once λ is obtained, μ can be found from

$$\mu = \Gamma^H \Lambda^\dagger x_2. \tag{5.30}$$

Remark. Note that there exists a simultaneous sign ambiguity (SSA) in the estimated results due to step (5.26); i.e., either $\{\lambda, \mu\}$ or $\{-\lambda, -\mu\}$ is found. Nonetheless, the SSA does not affect the data detection when we reconstruct $b_i(n)$'s. A similar observation is also made in Chap. 4.

2. *Frequency-Domain Approach*: Let $\tilde{\lambda}$ be the Z-point discrete Fourier transform (DFT) of λ with $Z \geq Q + 1$, whose mth entry is defined as

$$\tilde{\lambda}_m = \sum_{q=0}^{Q} \lambda_q e^{-j2\pi qm/Z}, \quad m = 0, \ldots, Z - 1. \tag{5.31}$$

On the other side, the mth element of the Z-point DFT of $\Gamma \lambda$ is

$$\xi_m = \sum_{q=0}^{Q} \lambda_q e^{-j2\pi \frac{(q-Q/2)M}{N}} e^{-j2\pi qm/Z} = e^{j\frac{\pi QM}{N}} \sum_{q=0}^{Q} \lambda_q e^{-j2\pi q(\frac{ZM}{N}+m)}. \tag{5.32}$$

If $R \triangleq \frac{ZM}{N}$ is an integer, then (5.32) becomes $e^{j\frac{\pi QM}{N}} \tilde{\lambda}_{\langle m+R \rangle_Z}$, where $\langle \cdot \rangle_Z$ denotes the modulo-Z operation. Then the mth element of the Z-point DFT of $x_1(m)$ is

$$\tilde{x}_1(m) = \xi_m \tilde{\lambda}_m = e^{j\frac{\pi QM}{N}} \tilde{\lambda}_m \tilde{\lambda}_{\langle m+R \rangle_Z}. \tag{5.33}$$

Our target is to retrieve Z unknown $\tilde{\lambda}_m$'s, $m = 0, 1, \ldots, Z - 1$ from Z equations

$$\tilde{\lambda}_m \tilde{\lambda}_{\langle m+R \rangle_Z} = \tilde{x}_1(m) e^{-j\frac{\pi QM}{N}} \triangleq c_m, \quad \forall m = 0, \ldots, Z - 1, \tag{5.34}$$

where c_m is defined as the corresponding constant.